ELECTROTHERMAL ANALYSIS OF VLSI SYSTEMS

ELECTROTHERMAL ANALYSIS OF VLSI SYSTEMS

Yi-Kan Cheng
Motorola, Inc.

Ching-Han Tsai
University of Illinois at Urbana-Champaign

Chin-Chi Teng
Silicon Perspective Corporation

Sung-Mo (Steve) Kang
University of Illinois at Urbana-Champaign

SPRINGER SCIENCE+BUSINESS MEDIA, LLC

Library of Congress Cataloging-in-Publication Data

A C.I.P. Catalogue record for this book is available
from the Library of Congress.

ISBN 978-1-4757-7373-6 ISBN 978-0-306-47024-0 (eBook)
DOI 10.1007/978-0-306-47024-0

Originally published by Kluwer Academic Publishers in 2000
Softcover reprint of the hardcover 1st edition 2000

Printed on acid-free paper.

Contents

List of Figures

List of Tables

Preface

With increasing complexity of VLSI chips, the task of developing state-of-the-art VLSI systems has become a highly challenging multidisciplinary task. Although in early days of MOS technology, silicon compilation looked promising, it has become difficult to fully automate the entire design flow from high-level design to mask generation due to many difficult physical design problems, including timing closure, power constraint, crosstalk, signal integrity, testability and reliability issues. In particular, the conventional practice of treating reliability qualification as a backend process has become no longer acceptable in view of excessive cost for design iterations. Attempts are under way to include reliability verification in the design flow so that expensive design iterations due to reliability problems can be avoided.

With foresight Semiconductor Research Corporation has provided strong support on our research of "design for reliability" at the University of Illinois at Urbana-Champaign for over a decade. New models and CAD capabilities have been developed and transferred to industry to address some of the serious reliability problems such as electromigration (EM) in metallic interconnects and electrostatic discharge (ESD) damages to I/O pads. With increasing concerns for on-chip power dissipation due to high packing density and high-frequency operation, electrothermal analysis has become critically important for accurate assessment of thermally activated device and circuit failures, and for timing analysis.

In this book we have attempted to provide in-depth coverage of important subjects required for electrothermal analysis of MOS VLSI circuits in an orderly manner. The underlying principles of circuit models and simulation algorithms in reliability CAD tools such as ILLIADS-T and iTEM are described in detail. For verification of design tool capability, chip design and bench test results are presented for electrothermal analysis of ring oscillators operating under digitally controlled thermal environment. We also present a "thermally skewed timing failure" phenomenon with detailed simulation result. This subject has not been discussed in any literature to our best knowledge, but such failures have been noticed by practicing VLSI design engineers.

It is our sincere desire that readers will find the contents of this book useful for practice and also for furtherance of research in this challenging field.

YI-KAN CHENG
CHING-HAN TSAI
CHIN-CHI TENG
SUNG-MO (STEVE) KANG

To ECE and CSL of the
University of Illinois at
Urbana-Champaign, and to
Semiconductor Research
Corporation

Acknowledgments

The authors wish to thank program managers of Semiconductor Research Corporation (SRC) and colleagues in semiconductor industry for their support, in particular Dr. Ralph Cavin, Dr. William Joyner, Dr. Justin Harlow of SRC, Dr. Charvaka Duvvury, Dr. P. B. Ghate of Texas Instruments, Dr. Ping Yang of Taiwan Semiconductor Manufacturing Company (TSMC), formerly of Texas Instruments, Dr. Shiuh-Wuu Lee of Intel Corporation. Authors also wish to acknowledge helpful discussions and encouragement from Profs. Elyse Rosenbaum, Timothy Trick, Ibrahim Hajj, Karl Hess, Ravi Iyer, Janak Patel of the University of Illinois at Urbana-Champaign, Profs. Chenming Hu, Ernest Kuh, Robert Brayton of the University of California at Berkeley. Prof. Ron Rohrer formerly of Carnegie-Mellon University, Prof. Stephen Director of the University of Michigan at Ann Arbor, and Dr. Herman Gummel of Lucent Technologies Ball Labs at Murray Hill have encouraged research on reliability-driven CAD. The Coordinated Science Laboratory under the directorship of Prof. W. Ken Jenkins and the Department of Electrical and Computer Engineering of the University of Illinois at Urbana-Champaign have provided excellent supports for research and preparation of this book.

Finally, the authors would like to thank their parents - Dong-Pyng and Shiaw-Chen Cheng, Hsiao-Lang and Mei-Hua Tsai, Yuan-Sun and Mei-Chu Teng - their wives - Hui-Chun (Angie) Cheng, Pei-Tzu Teng, and Myoung A (Mia) Kang - Ching-Han's friend - Kathy Chang - and their children - Jennifer and Jeffrey Kang - for their understanding and support during the writing of this book. Their love, patience and encouragement made this project possible. The authors express the deepest gratitude to them.

Foreword

Continuing increases in the levels of circuit integration and concomitant increases in performance are sustaining the trend of increasing power dissipation in VLSI systems. A consequence is that the impact of temperature on the successful operation and reliability of devices must be comprehended during the design process. For the past decade, the authors have led an effort to provide a framework, accompanied by tools, for the electrothermal analysis and design of integrated circuits and systems. This is a challenging field driven by the enormous complexity of integrated circuits and by the need for tractable, predictive, and executable models of electrical and thermal interaction physics.

This text provides a comprehensive formulation of the electrothermal analysis problem beginning with a summary of the sources of power dissipation in CMOS circuits and followed by a formulation of the effect of temperature on MOS devices. A general framework for thermal simulation of integrated circuits and packages is presented and then the fast timing electrothermal simulator, ILLIADS-T, is described. Applications include the study of temperature dependent electromigration reliability, captured in the simulator iTEM, and the placement of cells so as to mitigate temperature effects. The text concludes with the description of a methodology to predict the effects of temperature on the timing of integrated circuits.

The tools and methods described herein are finding widespread use in industrial applications by SRC members. We at the SRC are pleased to acknowledge the important contributions that have been made by the authors and expect that readers who are involved in electrothermal modeling will find the integrated perspective of this text to be very useful.

Dr. Ralph K. Cavin, Vice President
Semiconductor Research Corporation
February 2000

I

THE BUILDING BLOCKS

Chapter 1

INTRODUCTION

When the chip integration level increases and the device feature size decreases, the die yield goes down in most cases. Furthermore, the overall chip performance degradation can be significant due to parasitic effects and the associated reliability problems. Consequently, the chip *reliability* and chip *performance* have become equally important in high-performance very-large-scale-integrated (VLSI) system design.

The commonly considered reliability issues in a VLSI system are: hot carrier induced degradation, oxide breakdown, electrostatic discharge (ESD), electrical overstress (EOS), and electromigration (EM). Most of these issues have been discussed in detail in many introductory or advanced books. This book is intended to address another emerging and important reliability problem in VLSI systems: *electrothermal analysis of reliability and circuit performance.*

The electrothermal problem has long been a major concern in analog circuit design because the bipolar circuits consume a large amount of power and have the potential thermal runaway problem. Since current VLSI systems mainly consist of MOS devices, the power consumption is comparatively low and the electrothermal problem is seemingly not a threat. Unfortunately, it is not true when the technology scaling continues to be the trend of VLSI system design.

In the following, the electrothermal phenomena in a VLSI system are described. An overview of the generic electrothermal analysis flow is presented. The existing electrothermal analysis methods are also reviewed. Finally, the organization of this book is given.

Table 1.1. Trends of future microprocessor characteristics (ref: 1997 NTRS). Here the on-chip temperature is calculated assuming an ambient temperature of 75 °C.

Year	1999	2001	2003	2006	2009	2012
Tech. generation (nm)	180	150	130	100	70	50
Chip Size (mm^2)	340	385	430	520	620	750
Maximum Power (W)	90	110	130	160	170	175
Avg. Temperature (°C)	104	106	107	108	105	100

1.1. ELECTROTHERMAL PHENOMENA IN VLSI SYSTEMS

Due to the increasing packing density, higher operating speed, and larger scale of integration, the power density and on-chip temperature in integrated circuits continue to increase. For instance, the trend of future microprocessor characteristics is depicted in Table 1.1, which is extracted from the 1997 National Technology Roadmap for Semiconductors (NTRS). Table 1.1 shows the projection of the maximum power and the size of the chip. The operating temperatures are estimated by using the following formula:

$$T_i = T_a + P_{total} \times R^{th} \simeq T_a + \frac{P_{total}}{10^5 \times die\_size} \tag{1.1}$$

where T_i is the internal average chip temperature, T_a is the ambient temperature, P_{total} is the total power consumption of the design, and R^{th} is the equivalent thermal resistance of the packaging components (°C/W). The on-chip temperature of the packaged VLSI circuit not only can reach as high as 100 °C on average, but also can vary by as much as a few tens of degrees from one location to another.

Because the failure rate of microelectronic devices depends heavily on the localized operating temperature, hot spots due to high local-power dissipation have become a long-term integrated-circuit (IC) reliability concern in diverse applications such as high-performance microprocessors and digital signal-processing chips. Because of the complexity of a VLSI chip, the verification of chip performance at various operating temperatures relies heavily on computer simulations. Once the temperature profile is determined, several important issues shown in Fig. 1.1 can be addressed. It is clear that the thermal engineering can be used not only for reliability checking, but also as an additional degree of freedom for enhancing the circuit performance.

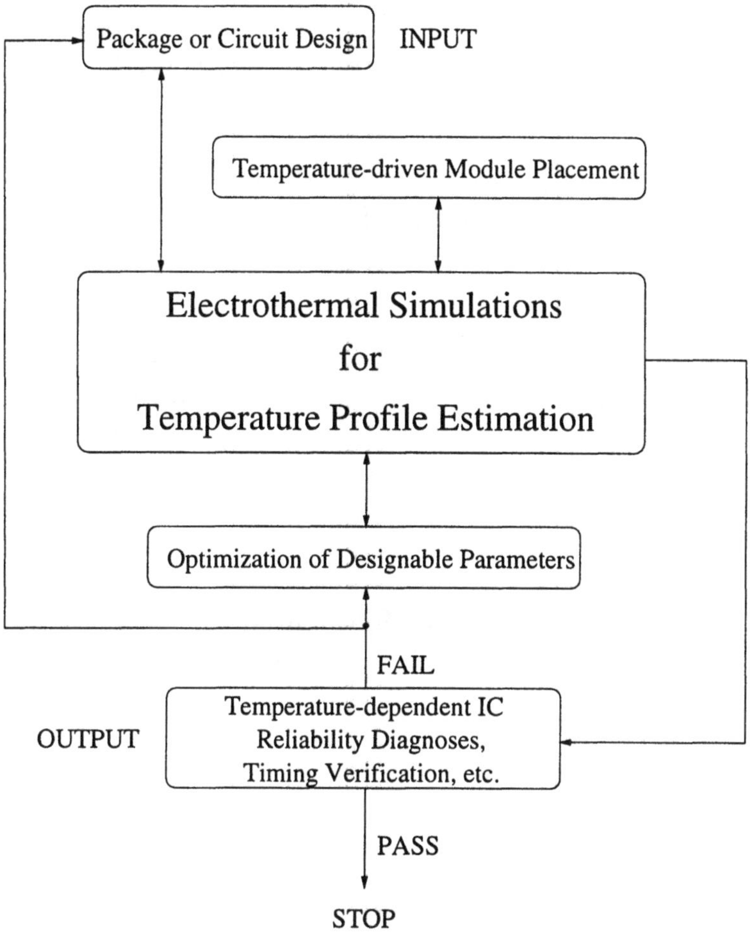

Figure 1.1. Applications of electrothermal CAD tools.

1.2. INTRODUCTION TO ELECTROTHERMAL SIMULATION

Electrothermal simulation consists of electrical and thermal simulations. The purpose of electrical simulation is to obtain the information on power dissipation and the performance of devices or circuits. On the other hand, the thermal simulation is used to find the temperature profile and to update all the temperature-dependent physical parameters of the the device or circuit model. This is illustrated in Fig. 1.2. The loop in Fig. 1.2 forms the basic mechanism of electrothermal simulation. The electrical and thermal relationships must be self-consistent for the system to remain stable. Otherwise the thermal runaway effects may occur.

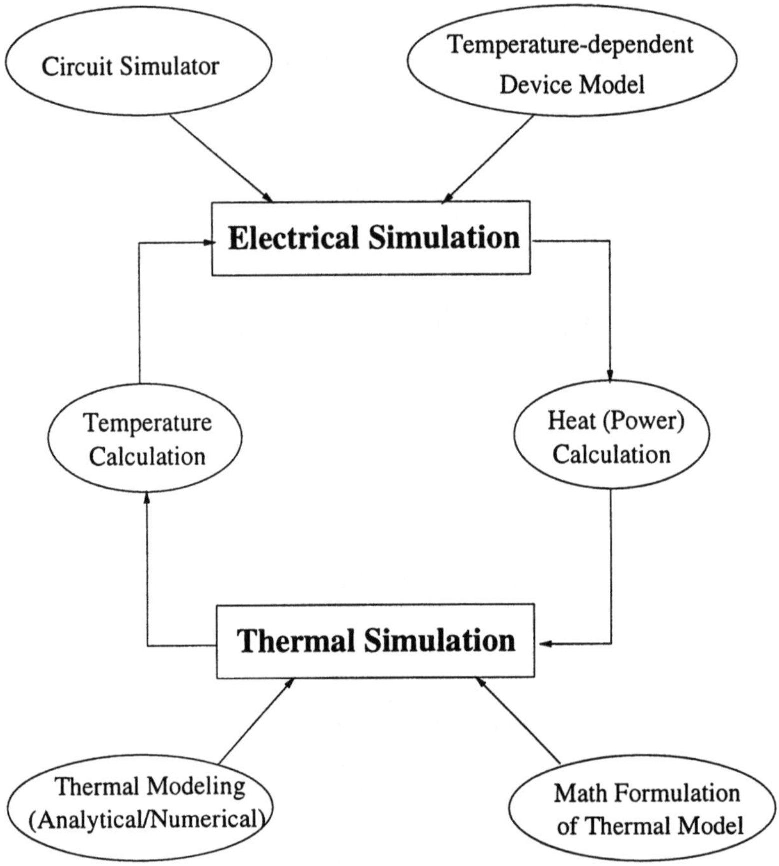

Figure 1.2. Elements of electrothermal simulations.

1.2.1 OVERVIEW OF ELECTROTHERMAL SIMULATION FOR ICS

Fukahori and Gray [1] comprehensively addressed the simulation of ICs in the presence of electrothermal interaction. Their focus was on the analog circuits where thermal feedback can severely degrade the circuit performance and distort the voltage transfer characteristics.

The electrothermal simulation procedure in [1] is illustrated in Fig. 1.3. A coupled set of nonlinear electrothermal equations is first generated. Next, those equations are represented by a matrix form and then linearized and solved by using the Newton-Raphson method. The linearized circuit matrix contains three parts:

1. Elements corresponding to the electrical circuit (Y_V)

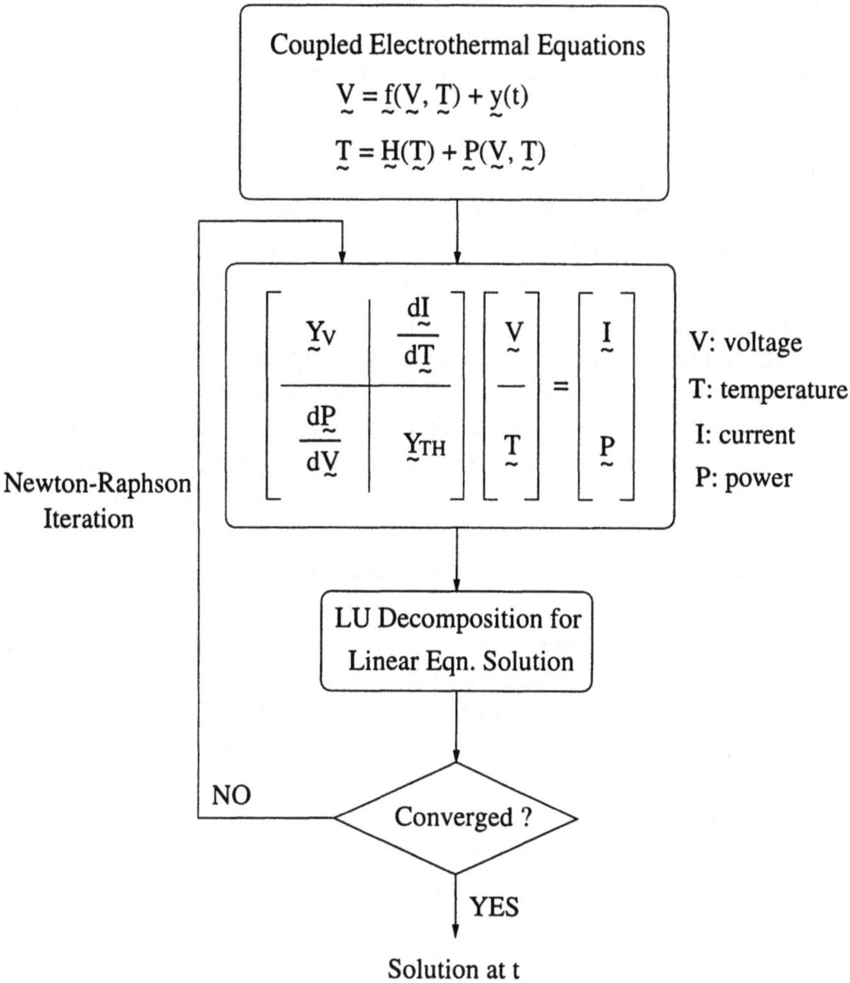

Figure 1.3. Electrothermal simulation procedure in [1].

2. Elements corresponding to the thermal circuit (Y_{th})

3. Elements corresponding to the coupling between the two circuits

The thermal circuit was generated by using the finite-difference method (FDM) for the simplified die-header structure. Elements corresponding to the coupling between the two circuits are the thermally controlled current sources corresponding to the temperature effects on the electrical physical parameters, and the electrically controlled power sources corresponding to the power dependence of the node voltages.

Once the matrix is solved and the dc solution is found at time t, as illustrated in Fig. 1.3, the transient solution of the temperature and the node voltage can be found by utilizing the preferred integration formula. In [1], the trapezoidal integration technique was employed. The above procedures are similar to those used in the circuit simulation programs such as SPICE [2]. The electrothermal simulator in [1] was applied to several analog circuits for the prediction of electrothermal interactions, both in dc transfer characteristics and in transient response. It was also pointed out that the simulation time was typically a factor of ten greater than the case when only the electrical effects were considered.

In 1982, a transistor-level electrothermal simulator was developed by Latif et al. [3]. It aimed at finding the temperature-dependent behaviors of power bipolar transistors. Each simulated device was partitioned into $n \times m$ (in 2-D case) sections connected in parallel by appropriate base resistors, where each section operated at its own temperature. A temperature-dependent Ebers-Moll model was used for each section. This model included the effects of avalanche multiplication, basewidth modulation, and current gain variations. The thermal network of the device was generated by using the 3-D finite-difference approach.

Two numerical techniques were proposed in [3] to solve the coupled electrothermal circuits. The first one was called the *direct method*, which is similar to the method proposed earlier in [1]. The second technique was called the *relaxation method.* This method divided the original problem into electrical and thermal systems. They were solved separately and the solutions were obtained by applying successive relaxation between the two systems. Both techniques have their own advantages and disadvantages. The direct method is more general and powerful for analyzing different problems such as dc, transient, and dc transfer characteristics. However, it is computationally more expensive and may not be able to handle all nonlinearities of the system. The relaxation method is more efficient, but convergence problems can occur under some biasing conditions.

Lee et al. developed a coupled electrothermal simulator for ICs in 1993 [4]. Its purpose was similar to that of [1], but with the focus on improving the simulation efficiency while preserving the accuracy. For dc analysis, the *incomplete Cholesky conjugate gradient* (ICCG) method[5] was used. For the transient analysis, the macromodeling method based on *asymptotic waveform evaluation* (AWE) [6] was employed.

The ICCG method is one of the relaxation methods that does not require the expensive LU factorization process to solve the network matrix as in the direct method. Combining incomplete Cholesky decomposition and conjugate gradient optimization, the ICCG method is known to be very efficient in solving symmetric and diagonally dominant systems such as 3-D interconnect structures or 3-D thermal networks. Simulation results for a 741 operational amplifier showed that the CPU time saved was 93% by using the ICCG method compared to the direct method [4]. More CPU and memory savings are expected for larger circuits.

AWE is a technique to find the time-domain response of a linear system by utilizing a reduced set of approximate poles and residues in the frequency-domain transfer function. These poles and residues are determined by applying a moment-matching method such as the *Padé approximation* [7]. The manner in which moments for a linear system are calculated is to successively perform the dc analysis of the system. For example, consider the RC circuit in Fig. 1.4(a). The first set of moments of the circuit is found by transforming the circuit in Fig. 1.4(a) into Fig. 1.4(b), replacing capacitors with zero-valued constant-current sources, and calculating the voltages across the current sources. The voltages m_{C1}, m_{C2}, and m_{C3} in Fig. 1.4(b) are the resulting first set of moments. The successive generations of higher-order moments are accomplished by setting the driver to zero and replacing each current source with the product of its previous moment and capacitance value. For illustration, the second set of moments for the circuit in Fig. 1.4(a) is found as shown in Fig. 1.4(c). Once the poles and residues are found by moment matching, transient response of the system can be subsequently calculated.

A linear thermal system can always be described in terms of the state equations in Eq. (1.2),

$$\begin{cases} \dot{x} = Ax + Bu \\ y = Cx + Du \end{cases} \tag{1.2}$$

where x is the state vector, u is the input vector, y is the output vector, and D is the vector related to the electrothermal coupling. Therefore, the AWE technique can be directly applied to this system to obtain the transient temperature response, which is computationally much more efficient when compared to a conventional time-domain integration method such as in SPICE. The transient electrothermal simulation was performed on the 741 operational amplifier, and the CPU time saved by using the AWE technique was about 85% in comparison to the trapezoidal integration method [4]. Transient simulations were done by Lee et al. for both bulk silicon and silicon-on-insulator (SOI) technologies, and the comparison of thermal effects between the two technologies was made.

A new circuit-level electrothermal simulator, iETSIM, was introduced by Díaz et al. in 1994 [8]. It simulates the transient electrothermal effects, with an emphasis on the electrical overstress (EOS) and electrostatic discharge (ESD) applications. ESD is one of the most prevalent causes for IC failures due to the short-duration high-current stress. Under such a stress, the breakdown of a device can occur. Because the second breakdown is thermally originated, electrothermal simulation is essential for an accurate ESD-induced failure analysis.

iETSIM is a coupled transient electrothermal simulator. To find the node voltages and circuit temperatures, a set of coupled electrothermal equations

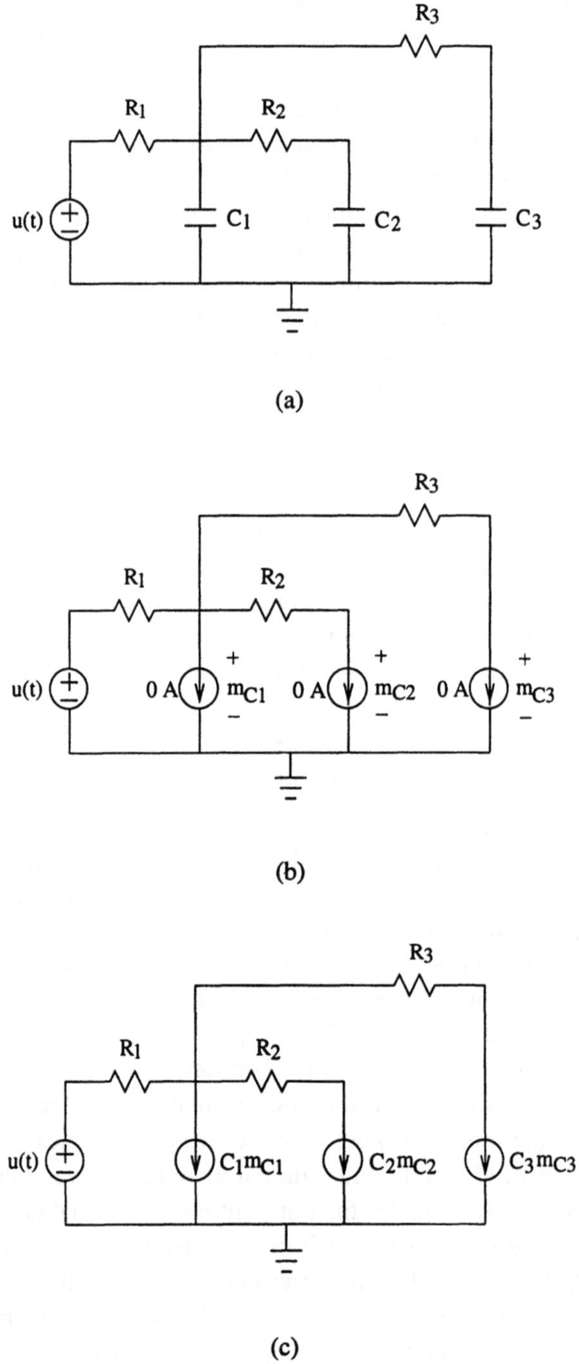

Figure 1.4. (a) An RC circuit example, (b) the dc circuit for the first-moment generation, and (c) the dc circuit for finding the second moment.

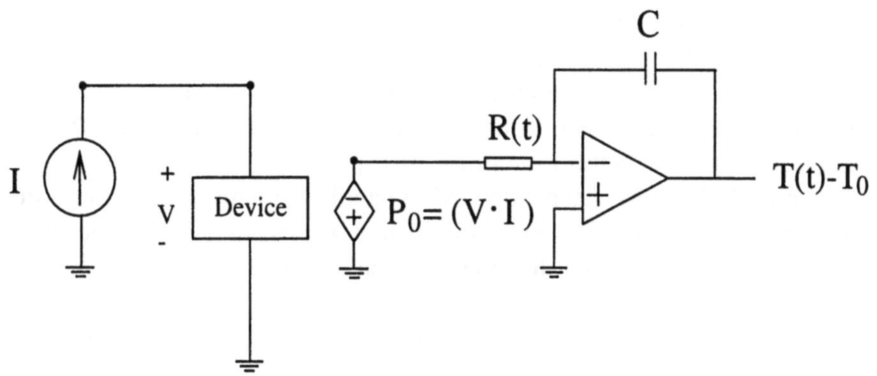

Figure 1.5. The integrator circuit used to implement the solution of the 3-D heat diffusion equation.

is formed and solved by using the standard *modified nodal analysis* (MNA) technique as shown in Fig. 1.3. For the electrical part, a new model and algorithm for avalanche breakdown were developed for accurate ESD/EOS simulation. This new algorithm was shown to be much simpler, more robust and more efficient than the algorithms introduced earlier in [9]. For the thermal part, a novel temperature model based on an electrical analog implementation of the time-dependent 3-D heat-diffusion equation was developed. It employed the solution of the 3-D heat-diffusion equation derived by Dwyer et al. [10]. For a heat source with dimensions $a \times b \times c$ and a constant power value P_0, the transient temperature distribution due to this source can be written as [10]

$$T(\vec{r}, t) = T_0 + \frac{P_0}{\rho C_p abc} \int_0^t G(x, a, \tau) G(y, b, \tau) G(z, c, \tau) d\tau. \tag{1.3}$$

In Eq. (1.3), $\vec{r}$ is the location of the observation point with respect to the center of the heat source, T_0 is the ambient temperature, ρ is the mass density, C_p is the specific heat, and $G(x, a, \tau)$, $G(y, b, \tau)$, and $G(z, c, \tau)$ are the Green's functions.

In iETSIM, the integral over time in Eq. (1.3) is evaluated by using an electrical equivalent integrator circuit shown in Fig. 1.5. In this circuit, a power monitor (P_0) and a time-dependent resistor (R) are provided to convert power to the temperature rise above the ambient temperature T_0. The time-dependent resistor can be obtained from Eq. (1.3) and is given as

$$R(\vec{r}, t) = \frac{\rho C_p abc}{C G(x, a, t) G(y, b, t) G(z, c, t)} \tag{1.4}$$

where C is chosen so that the matrix entries become more even, and its typical value is 1 μF.

With the implementation in Fig. 1.5, iETSIM is more efficient than the relaxation method in [3], especially when the circuit contains more than one device or when the temperature gradient is steep under ESD stress. In order to simulate the complete coupling between various heat sources in an ESD protection circuit, the current summation property of the integrator can be used as suggested by the superposition principle [8].

Recently iETSIM has been extended to handle the temperature calculation of devices with time-varying power dissipation values. In this case, the device temperature is given by the following convolution equation:

$$T(\vec{r}, t) = T_0 + \frac{1}{\rho C_p abc} \int_0^t P(t - \tau) G(x, a, t) G(y, b, t) G(z, c, t) d\tau. \quad (1.5)$$

Numerically computing the convolution equation in Eq. (1.5) is expensive. For the sake of efficiency, regionwise exponential (RWE) approximation is applied to the Green's functions in Eq. (1.5), in order to perform the convolution recursively for temperature calculation. Readers may refer to [11] for further details.

1.3. ILLIADS-T: AN ELECTROTHERMAL SIMULATOR FOR VLSI SYSTEMS

A fast-timing simulation based electrothermal simulator, called ILLIADS-T, was developed [12]. ILLIADS-T was designed to simulate the digital VLSI circuits. The flowchart of ILLIADS-T simulation procedure is shown in Fig. 1.6. The main features of ILLIADS-T are listed below.

1. To achieve the computational time efficiency required by large circuits, ILLIADS-T uses a fast-timing simulator, ILLIADS (ILLInois Analogous Digital Simulator) [13], to calculate the power dissipated by each *logic gate*. Each gate is then viewed as a heat source in thermal simulation. ILLIADS has the following advantages:

   - The speedup of ILLIADS over SPICE-like programs increases linearly with the circuit size as measured in terms of the transistor count
   - The speedup can be further enhanced by introducing the *incremental* electrothermal simulation technique [14]
   - an accurate temperature-dependent modeling method for the MOS device was developed based on the regionwise quadratic (RWQ) modeling technique [15]. With this method, the accuracy of delay and power values estimated by ILLIADS is comparable to SPICE for a wide range of temperatures (27 °C - 120 °C).

2. The coupled electrothermal simulation methods such as those introduced earlier are time consuming. The total simulation time is first divided into

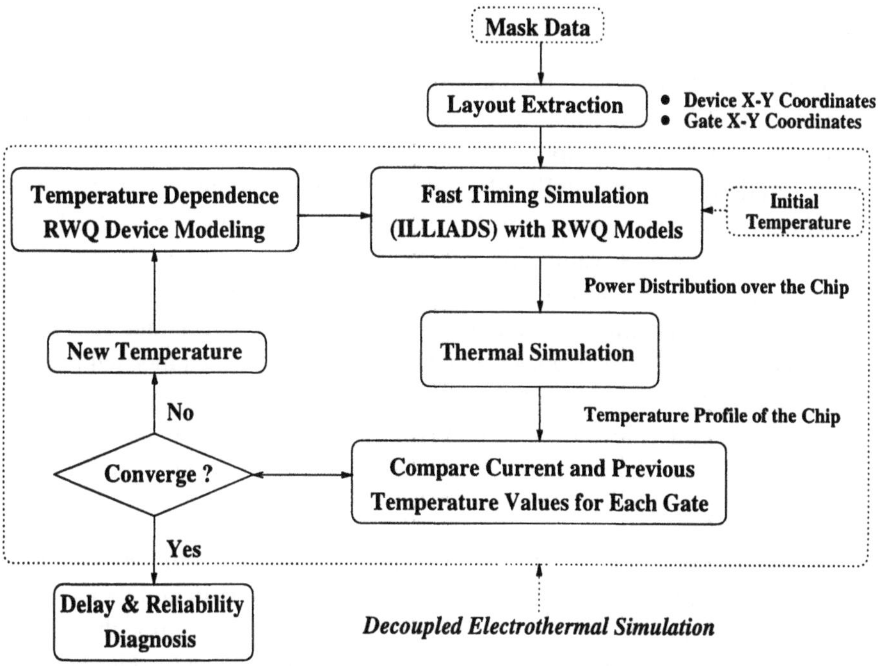

Figure 1.6. Flowchart of ILLIADS-T electrothermal simulation.

many small time intervals, then the power and temperature values are updated and coupled for each time interval. This kind of approach is ideal only for transient simulation on small circuits. ILLIADS-T, which is designed to find the chip-level steady-state temperature distribution and the resulting circuit performance, uses a much more feasible approach for VLSI circuits. It starts with an initial guess of the average chip temperature and then calculates the average power for each gate based on the current waveform drawn from the power supply. Next, the gate power values are fed to the thermal simulator to estimate the temperature profile. The temperature profile is then used to update the device model parameters for the second round of power calculation. This process continues until convergence is obtained and the steady-state temperature profile is found.

The above approach *decouples* the power and temperature calculation. The decoupling strategy is justified by the fact that the time required for the on-chip temperature to reach steady state (i.e., thermal time constant) is several orders of magnitude longer than the clock signal period (i.e., electrical time constant) in digital circuits [16]. In other words, the chip temperature does not immediately follow the instantaneous power dissipation, and thus the average power instead of the instantaneous power is used in the steady-state temperature calculation in ILLIADS-T.

3. Previous electrothermal simulators were developed mainly for the temperature profile estimation of SSI or MSI circuits [1, 4, 8], therefore the thermal boundary conditions were simplified. Moreover, the 1-D/2-D thermal simulations were usually adopted. For VLSI/ULSI chips with complex packaging structures, the simplified boundary conditions and 1-D/2-D approaches may not be valid. To handle this problem, a thermal simulation framework, iTEMP, has been built in ILLIADS-T to solve the 3-D heat equations for the chip substrate and to model the packages and heat sinks as *effective thermal resistances*. iTEMP can handle various thermal boundary conditions at any side of the chip with no limitations. A hierarchical approach was also developed in this thermal simulation framework in order to quickly identify the on-chip hot spots and to subsequently pinpoint the hot-spot temperatures.

4. The chip temperature can be found with full top-down automation. Once the chip dimensions, packaging materials, device I-V data and thermal parameters are specified by the user, ILLIADS-T requires only the layout description file (e.g., CIF or GDSII format) to find the steady-state temperature profile and the corresponding circuit performance and reliability.

5. By using the RWQ modeling technique instead of the complex MOS models as in [17], temperature-dependent power and delay estimation can be done in ILLIADS-T even when only measured data are available and the MOS models have not been fully developed or characterized. This makes ILLIADS-T device-model-independent, and thus applicable to the advanced CMOS technologies.

Referring back to Fig. 1.6, the primary input to ILLIADS-T is the layout description file of the target VLSI chip. A layout extractor has been developed to obtain the electrical circuit that the layout represents, as well as to identify the location of each device. A standard device specification in the netlist generated by the layout extractor in ILLIADS-T is shown below:

MOS_name ND NG NS NB MODEL_name ⟨L=VAL⟩ ⟨W=VAL⟩ ⟨AD=VAL⟩ ⟨AS=VAL⟩ ⟨PD=VAL⟩ ⟨PS=VAL⟩ XMIN YMIN XMAX YMAX

where XMIN, YMIN, XMAX, and YMAX define the bounding box of a MOS device layout, and MODEL_name specifies a particular RWQ model for a MOS device.

ILLIADS-T then calculates the bounds of each logic gate according to the coordinates of the bounding boxes of MOS devices within this gate. Next, the average power dissipation from each gate at the initial temperature is calculated by ILLIADS. iTEMP will take as input the power values and the coordinates of heat sources to calculate the on-chip temperature profile by solving the heat equations. In particular, the average temperature of each gate is found. At this stage, each gate has its updated local temperature and ILLIADS must be rerun to find the new average power values under the new temperature distribution. This iterative procedure stops when the updated temperature of each gate no longer has any significant change from the previous value. Empirical results shown in [12] indicate that this process is efficient and usually converges within two or three iterations. Note that in CMOS circuits, the short-circuit power can account for approximately 25% of the total IC power consumption [18]. The temperature-induced variations of the short-circuit power and/or the switching activity are what necessitate a few iterations during ILLIADS-T simulation.

1.4. OVERVIEW OF THIS BOOK

This book addresses the issues related to electrothermal problems in modern VLSI designs from the modeling and simulation perspectives. It is intended to cover the most important electrothermal reliability and performance issues that can be encountered in VLSI system design. Solid-state transistors, interconnects, logic gates, macros, chips, and packages, are all temperature-sensitive VLSI design objects that will be covered in this book.

The first few chapters are designed to present the fundamental building blocks in an electrothermal simulation environment. Chapter 2 discusses the power analysis methods. As shown in Fig. 1.2, power analysis is used to determine the amount of power dissipation to be used in thermal analysis. Therefore, it is the very first building block in electrothermal analysis. Chapter 2 starts with the introduction of sources of power consumption in a CMOS circuit, followed by three different power analysis techniques. These three techniques have their own advantages and disadvantages, which will be addressed and compared in detail.

In Chapter 3, the temperature-dependent MOS device modeling is presented. The temperature-dependent modeling of the threshold voltage and the channel carrier mobility of a MOS transistor is given. The three scattering mechanisms that determine the carrier mobility in the solid are also described. Two temperature-dependent MOS device models are presented. The first one is the BSIM model, which is based on the sensitivity analysis of the sensitive parameters in the original BSIM model. The second one is the RWQ model, which is based on the regionwise fitting to the experimental data and the inclusion of the scattering mechanisms in mobility modeling.

Chapter 4 concentrates on thermal analysis for VLSI systems. This chapter begins with the introduction of the heat equation and the common thermal boundary conditions. A complete thermal simulation framework is presented.

This framework includes a fast thermal simulator, a numerical thermal simulator, an analytical thermal simulator, and a package thermal simulator. The mathematical formulations of the solutions to these thermal simulation methods are discussed. Examples of simulating the VLSI system using these methods are also given.

Chapter 5 focuses on the discussion of the fast-timing electrothermal simulation. A fast timing simulator developed at the University of Illinois at Urbana-Champaign, ILLIADS, is first presented. A fast-timing electrothermal simulator, ILLIADS-T, was developed by combining ILLIADS with the thermal simulation framework discussed in Chapter 4. An incremental simulation strategy used in ILLIADS-T is illustrated in this chapter. A tester chip was designed for verifying the accuracy of ILLIADS-T. The details of the tester chip design, experimental setup, tester chip calibration, and chip temperature measurement, are presented. The experimental results are compared with the ILLIADS-T simulation results. Finally, a number of circuits are simulated, and the impact of the thermal effect on the circuit performance is examined.

The last three chapters in this book address three important applications of the electrothermal analysis, based upon the building blocks discussed in previous chapters. In Chapter 6, the temperature-dependent electromigration diagnosis method is presented. The chapter first illustrates how significantly the interconnect temperature affects its electromigration mean time to failure, followed by the introduction of the electromigration phenomena. The dependence of the electromigration lifetime on the current density, current waveforms, and the metal length and width, is described. Based one the dependence, a electromigration model suitable for circuit simulation is presented. Next, the overview of existing electromigration analysis methods is given. Finally, a temperature-dependent electromigration diagnosis tool developed at the University of Illinois at Urbana-Champaign, called iTEM, is discussed. A lumped thermal model used to find the temperatures of the multilayered interconnect system is developed and discussed. Finally, a number of circuits are simulated using iTEM. The electromigration mean time to failure is predicted by iTEM, and the importance of including interconnect temperature in the analysis is demonstrated.

Chapter 7 addresses the issues related to temperature-driven cell placement for uniform substrate thermal distribution. Two approaches for deriving the compact substrate thermal model are illustrated. The first approach employs the superposition principle to construct a transfer thermal resistance matrix, and the second approach involves the direct manipulation of the nodal matrix equation. The comparison of runtime efficiency of these two approaches is given. Two thermal placement algorithms, one suitable for the standard-cell placement and the other suitable for the macrocell placement, are presented and discussed. The algorithms have been tested on many benchmark circuits.

The thermal placement results show that, in general, the temperature profile is improved at the cost of longer simulation time. The total wire length and area after thermal placement are also compared with those generated by the conventional placement algorithm (i.e., no temperature is considered).

In Chapter 8, an integrated framework for temperature-driven power and timing analysis is presented. Since power and timing are closely related, they are treated together in this chapter. The relationship between power, temperature, and timing will be explained. An overview of timing analysis is given in order to show its importance to VLSI system design. Two timing analysis methods are introduced: dynamic method and static method. Dynamic timing analysis uses user-specified input patterns to simulate the circuit delay. It is the most accurate way of predicting path timing. Static timing analysis is conceptually very different from the dynamic analysis in that no input patterns are required. The false-path problem associated with the static timing analysis is examined and different solutions are provided. Two distinct approaches in static timing analysis, path-oriented approach and block-oriented approach, are presented. Because the block-oriented approach is extremely efficient and widely used in high-performance VLSI design, it will be the focus of our discussion. The step-by-step illustration of how to calculate the arrival times, the required times, and the slacks in the block-oriented approach is given. All of the above timing analysis methods will be compared. Next, the delay modeling for timing analysis is discussed. The temperature dependence of the gate and interconnect delay will also be addressed. A statistical technique for estimating average power of each logic gate in the circuit is presented. The estimated average power are used to find the nominal on-chip temperature distribution by utilizing a power-temperature iteration scheme. This two-level iteration scheme will be described. Finally, the experimental results are demonstrated. The nominal temperatures are statistically estimated and the timings of the benchmark circuits are found by using both dynamic and static timing analysis methods. It will be shown that the on-chip temperature rise and temperature gradient can cause different critical path and critical timing in comparison to the case where a uniform temperature distribution is assumed.

References

[1] K. Fukahori and P. R. Gray, "Computer simulation of integrated circuits in the presence of electrothermal interaction," *IEEE Journal of Solid-State Circuits*, vol. 11, pp. 834–846, Dec. 1976.

[2] L. W. Nagel, *SPICE2: A Computer Program to Simulate Semiconductor Circuits*. PhD thesis, Dept. of Electrical Engineering, Univ. of California at Berkeley, 1975.

[3] M. Latif and P. R. Bryant, "Network analysis approach to multidimensional modeling of transistors including thermal effects," *IEEE Transactions on Computer-Aided Design of Integrated Circuits and Systems*, pp. 94–101, Apr. 1982.

[4] S. S. Lee and D. J. Allstot, "Electrothermal simulation of integrated circuits," *IEEE Journal of Solid-State Circuits*, vol. 28, pp. 1283–1293, Dec. 1993.

[5] J. A. Meijerink and H. A. van der Vorst, *Mathematics of Computation*, vol. 31, pp. 148–162, 1977.

[6] L. T. Pillage and R. A. Rohrer, "Asymptotic waveform evaluation for timing analysis," *IEEE Transactions on Computer-Aided Design of Integrated Circuits and Systems*, vol. 9, pp. 352–366, Apr. 1990.

[7] G. A. Baker Jr., *Essentials of Padé Approximants.* New York, NY: Academic Press, 1975.

[8] C. H. Díaz, S. M. Kang, and C. Duvvury, "Circuit-level electrothermal simulation of electrical overstress failures in advanced MOS I/O protection devices," *IEEE Transactions on Computer-Aided Design of Integrated Circuits and Systems*, vol. 13, pp. 482–493, Apr. 1994.

[9] C. H. Díaz and S. M. Kang, "New algorithms for circuit simulation of device breakdown," *IEEE Transactions on Computer-Aided Design of Integrated Circuits and Systems*, vol. 11, pp. 1344–1354, Nov. 1992.

[10] V. Dwyer, A. Franklin, and D. Campbell, "Thermal failure in semiconductor devices," *Solid-State Electronics*, vol. 33, pp. 553–560, May 1990.

[11] T. Li, C. H. Tsai, and S. M. Kang, "Efficient transient electrothermal simulation of CMOS VLSI circuits under electrical overstress," in *Proceedings of the ACM/IEEE International Conference on Computer-Aided Design*, pp. 6–11, Nov. 1998.

[12] Y. K. Cheng, P. Raha., C. C. Teng, E. Rosenbaum, and S. M. Kang, "ILLIADS-T: An electrothermal timing simulator for temperature-sensitive reliability diagnosis of CMOS VLSI chips," *IEEE Transactions on Computer-Aided Design of Integrated Circuits and Systems*, vol. 17, pp. 668–681, Aug. 1998.

[13] Y. H. Shih, Y. Leblebici, and S. M. Kang, "ILLIADS: A fast timing and reliability simulator for digital MOS circuits," *IEEE Transactions on Computer-Aided Design of Integrated Circuits and Systems*, vol. 12, pp. 1387–1402, Sept. 1993.

[14] Y. K. Cheng and S. M. Kang, "Improvement on Chip-Level Electrothermal Simulator - ILLIADS-T," in *Proceedings of the IEEE International Symposium on Circuits and Systems*, May 1996.

[15] A. Dharchoudhury, S. M. Kang, K. H. Kim, and S. H. Lee, "Fast and accurate timing simulation with regionwise quadratic models of MOS I-V characteristics," in *Proceedings of the ACM/IEEE International Conference on Computer-Aided Design*, pp. 208–211, Nov. 1994.

[16] R. Darveaux, I. Turlik, L. T. Hwang, and A. Reisman, "Thermal stress analysis of a multichip package design," *IEEE Transactions on Components, Hybrids, and Manufacturing Technology*, vol. 12, pp. 663–672, Dec. 1989.

[17] C. P. Wan and B. J. Sheu, "Temperature dependence modeling for MOS VLSI circuit simulation," *IEEE Transactions on Computer-Aided Design of Integrated Circuits and Systems*, vol. 8, pp. 1065–1073, Oct. 1989.

[18] A. M. Hill, *Switching Density Analysis for Power and Reliability in VLSI Circuits*. PhD thesis, Dept. of Electrical and Computer Engineering, University of Illinois at Urbana-Champaign, 1996.

Chapter 2

POWER ANALYSIS FOR CMOS CIRCUITS

2.1. INTRODUCTION

To calculate the on-chip temperature profile, according to Fig. 1.2, the power distribution must be calculated first. Because of the increasing use of portable electronic applications such as cellular phone, laptop computers and personal digital assistant (PDA) devices, low power is the trend for modern processor design in order to reduce the chip temperature and to prolong the operation time between two battery charge-ups. Without proper thermal engineering, the overheating in VLSI chips can degrade the circuit performance and reduce the chip life time. For those high-power chips, temperature control must be done by using costly packaging materials and efficient heat-dissipating structures.

Because power management is important, *power analysis* has become indispensable for VLSI design and is one of fields that is currently under extensive investigation. A common goal of power analysis is to accurately and efficiently calculate the power consumption of the system under analysis. In this chapter, several power analysis methods will be described. First, the definition of power dissipation in CMOS circuits and the common sources of power consumption will be discussed in the following section.

2.2. SOURCES OF POWER CONSUMPTION IN CMOS TECHNOLOGY

A CMOS digital circuit always consumes power whether its logic state undergoes dynamic transitions or remains unchanged. Its power consumption is comprised of the following four components:

- dynamic (switching) power

- internal power

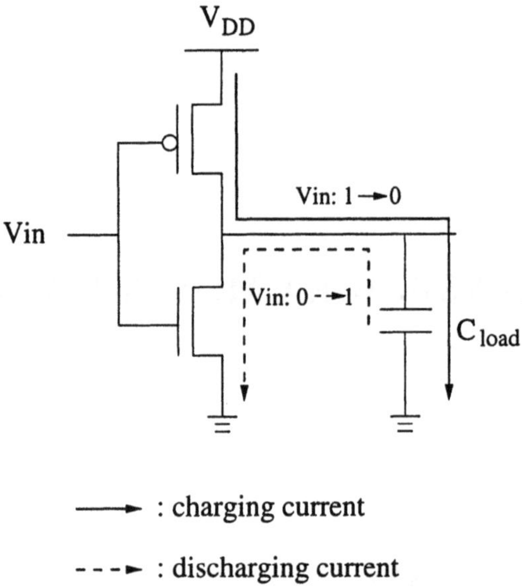

Figure 2.1. Illustration of dynamic power consumption in a CMOS inverter.

- short-circuit power
- leakage power

The cause of each component and the significance of its contribution to the total power consumption will be explained as follows.

2.2.1 DYNAMIC POWER

Dynamic power occurs when the output of the CMOS logic gate switches. During the switch, the output parasitic capacitances are either charged up to the supply voltage level or discharged to the ground level (See Fig. 2.1). During the charge-up phase, half of the energy supplied by the power source is stored in the output loading capacitance. The other half has been dissipated by the PMOS transistor. During the discharge phase, the remaining charge is removed from the capacitor and its energy is dissipated by the NMOS transistor. The average dynamic power consumption of a logic gate can be expressed as

$$P_{dyn} = 0.5 \times C_{load} \times V_{DD}^2 \times f_{clk} \times E(\text{transitions}) \qquad (2.1)$$

where V_{DD} is the power supply voltage, f_{clk} is the global clock frequency, $E(\text{transitions})$ is the expected number of transitions per clock cycle at the gate output, and

$$C_{load} = \sum_{i \in fanout} C_G^i + C_{wire}. \qquad (2.2)$$

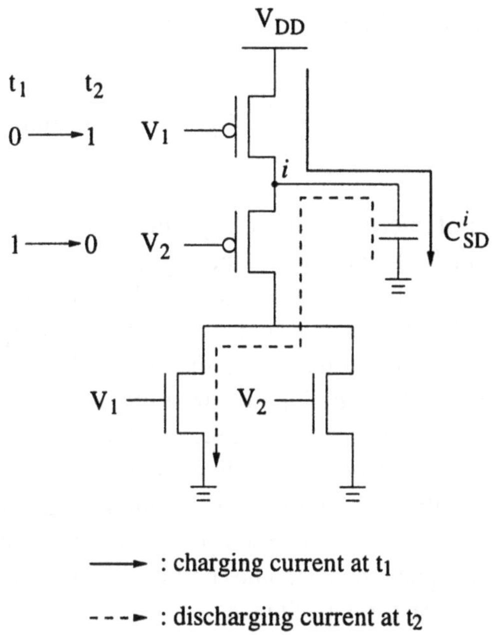

Figure 2.2. Charging and discharging of an internal node of 2-input NOR gate.

Here, C_G^i is the gate capacitance of the ith fanout and C_{wire} is the interconnect capacitance of the driven net.

Dynamic power is used for the logic evaluation by propagating the output states of logic gates. Therefore the dynamic power must be consumed in order to realize the functionality of a circuit. Given a processing technology and a functional description, there exists a theoretical *lower bound* of the power that must be consumed. This lower bound is determined by the amount of computation and is independent of the implementation [1, 2, 3, 4]. In a reasonably designed circuit, dynamic switching power usually accounts for the major portion of the total power consumption [5].

2.2.2 INTERNAL POWER

When the inputs of a logic gate are switching, it is possible that certain internal capacitances are charged or discharged without changing the output logic states. When this occurs, the *internal power* is consumed. One example is shown in Fig. 2.2 [6], where a two-input NOR gate is considered. If the inputs V_1 and V_2 are changing from 01 at t_1 to 10 at t_2, the output remains unchanged at state 0. However, after t_2, the capacitance at node i is discharged.

The internal power consumption happens at the internal nodes of the logic gate, therefore it cannot be captured by the dynamic switching power model in

Eq. (2.1). The internal power of a logic gate can be calculated as

$$P_{int} = 0.5 \times V_{DD}^2 \times f_{clk} \times \sum_{i=1}^{N_{in}} [C_{SD}^i \cdot E^i(\text{transitions})] \qquad (2.3)$$

where N_{in} is the number of internal nodes, C_{SD}^i is the source/drain capacitance of the internal node i, and $E^i(\text{transitions})$ is the expected number of transitions at node i. Note that Eq. (2.3) is very similar to Eq. (2.1) except that the power consumption is measured at the internal nodes instead of the output node.

From Eq. (2.3) it can be seen that when the source/drain capacitance is large or when the number of internal nodes in a circuit is big, the internal power consumption can be considerable. A recent study [6] showed that the internal power is on average $\sim$16% of the power consumption due to the gate capacitances. However, in the deep-submicron technologies, the internal power usually accounts for less than 5% of the total power consumption [5]. This is because the interconnect capacitance becomes dominant in the parasitic loading in comparison with the transistor capacitance.

2.2.3 SHORT-CIRCUIT POWER

Another source of power dissipation in a CMOS circuit is the direct flow of current from power source to ground. It is called the *short-circuit power*, which occurs when both the NMOS and PMOS transistors are conducting simultaneously. Such a path should never exist in a dynamic circuit because the precharge and evaluate transistors should never be on at the same time, or malfunction will occur.

Ideally, if a CMOS logic gate is driven by step input signals, either PMOS or NMOS transistors (but not both) will be conducting at a time. Unfortunately, input signals always have nonzero transition time because of the nonzero loading of the previous logic stage. Let us consider a CMOS inverter containing an NMOS transistor with threshold voltage $V_{T,n}$ and a PMOS transistor with threshold voltage $V_{T,p}$. During transition, when the voltage of the input signal V_{IN} satisfies

$$V_{T,n} < V_{IN} < V_{DD} - |V_{T,p}|, \qquad (2.4)$$

both transistors are on and the short-circuit power is consumed. The total amount of short-circuit power dissipation is a function of the on-time of the transistors and the operating modes of the devices.

Short-circuit power estimation has attracted many research interests in recent years [7, 8, 9, 10, 11]. Although it is difficult to derive an exact formula that is valid for all operating conditions in a circuit, simple expressions have been derived for some special cases. For instance, considering an *unloaded* inverter

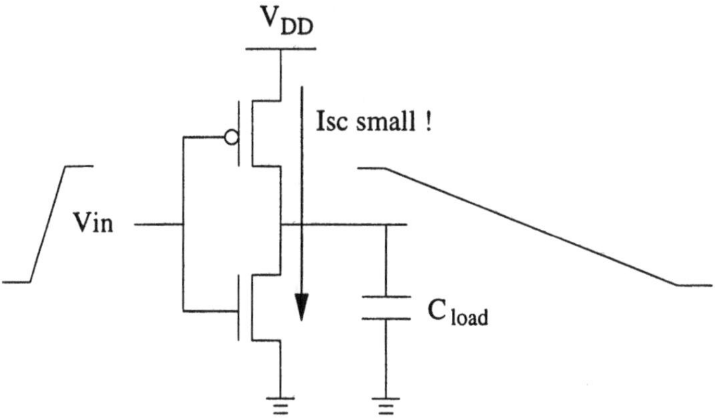

Figure 2.3. Short-circuit power for inverter with large load.

with $V_{T,n} = V_{T,p} = V_T$, the short-circuit power can be calculated as [7]:

$$P_{sc} = \frac{\beta}{12}(V_{DD} - 2V_T)^3 \cdot \tau \cdot f_{clk} \tag{2.5}$$

where β is the MOS transistor gain factor and τ is the input transition time. Note that the input transition time determines the length of the period during which Eq. (2.4) holds. The longer τ is, the more the short-circuit power is consumed.

Now let us qualitatively consider the impact of the loading capacitance on short-circuit current. In Fig. 2.3, the large output loading causes the output transition to be slow. Under such circumstances, the input signal moves through the transient period before the output starts to change. As a result, the PMOS is off and only a small amount of short-circuit current is carried. The opposite case is given in Fig. 2.4, where the small output loading causes immediate output transition. A considerable amount of short-circuit current will flow since the drain-source voltage of the PMOS transistor equals V_{DD} for most of the transition period.

From the above discussion, it can be concluded that the short-circuit power of a gate is minimized if the output rise/fall time is larger than the input rise/fall time. A common practice to minimize the short-circuit power of a circuit in a global way, however, is to match the rise/fall times of the input and output signals for every logic gate [7].

2.2.4 LEAKAGE POWER

All of the power components described above manifest themselves when a circuit is switching. Ideally there should be no power consumption if the circuit is in steady state. However, there is always a leakage current flowing

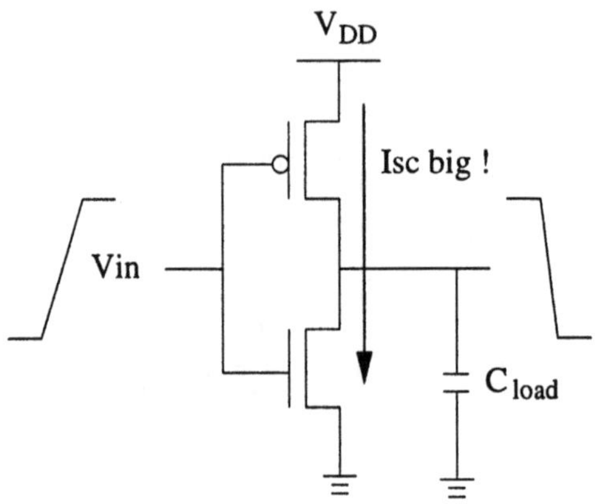

Figure 2.4. Short-circuit power for inverter with small load.

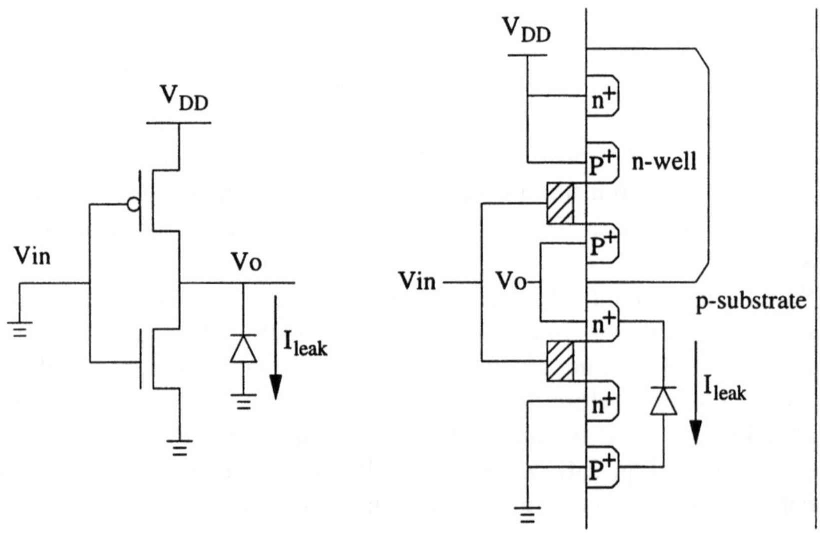

Figure 2.5. Leakage current at reversed-biased diode junction.

through the reverse-biased diode junctions of the transistors located between the source or drain and the substrate, as depicted in Fig. 2.5. The resulting power consumption is called the *diode leakage power*.

The magnitude of the diode leakage current can be expressed as

$$I_{dl} = A_D \times J_S \tag{2.6}$$

where A_D is the drain diffusion area and J_S is the leakage current density. Since the leakage current saturates at relatively small reverse bias potential, it is

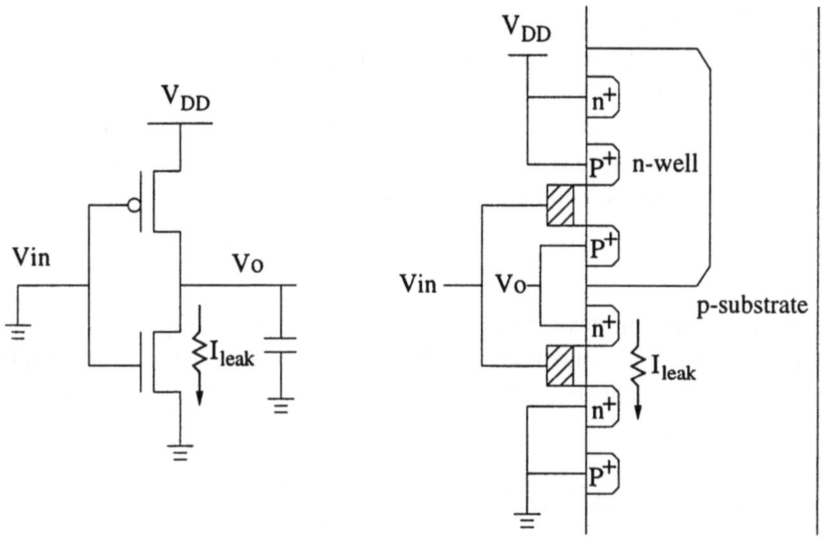

Figure 2.6. Subthreshold leakage current.

roughly independent of the supply voltage. Moreover, the diode leakage current is caused by the thermally generated carriers. Therefore, its value increases exponentially with the increasing junction temperature. Since the diode leakage current is generally small compared with other power components, it is often ignored in power estimation.

A more important source of leakage current is the subthreshold leakage current, and the resulting power consumption is called the *subthreshold leakage power*. An MOS transistor can experience a drain-source current even when the gate-source voltage is smaller than the threshold voltage, as shown in Fig. 2.6. The closer the threshold voltage is to zero volts, the greater the leakage current.

In the subthreshold regime, an MOS transistor behave similarly to a bipolar transistor, and its $I - V$ characteristics can be described by [12]:

$$I_{sl} = K \cdot e^{(V_{GS}-V_T)/(nV_t)}(1 - e^{-V_{DS}/V_t}) \tag{2.7}$$

where K is a function of the technology, V_t is the thermal voltage (KT/q), V_T is the threshold voltage and $n = 1 + \Omega t_{ox}/D$, where t_{ox} is the gate oxide thickness, D is the channel depletion width, and $\Omega = \epsilon_{si}/\epsilon_{ox}$. It can be seen that as V_T decreases, the magnitude of the subthreshold leakage current grows exponentially. In order to offset this effect, the threshold voltage of the MOS transistors is generally kept above a certain level (e.g., 0.5 V).

Technology scaling tends to lower the power supply voltage. In order to maintain or even increase the driving capability of the transistor current and thus the circuit speed, the threshold voltage needs to be scaled down as well. However, doing so increases the subthreshold leakage power, which is undesirable in the low power design. To solve the above dilemma, the *dual-*V_T technology is widely employed for the high-performance processor design. In such a design, the high-V_T devices are generally used for low power. For devices on the timing critical paths, however, the low-V_T technology is used.

For both sources of leakage (i.e., diode leakage and subthreshold leakage), their power dissipation can be expressed by

$$P_{leakage} = I_{leakage} \times V_{DD} \tag{2.8}$$

where $I_{leakage}$ is denotes either I_{dl} in Eq. (2.6) or I_{sl} in Eq. (2.7). Since the leakage power could contribute significantly to the total power consumption, several CAD tools have been developed for analyzing [13] and optimizing [14] the leakage power.

2.3. POWER ANALYSIS OVERVIEW

In general, a power analysis tool can be classified based on one of the following criteria:

- *Analysis Levels:* architectural level, register transfer level (RTL), gate level, and transistor level, etc.
- *Analysis Engines:* SPICE, fast-timing, switch-level, etc.
- *Analysis Techniques:* deterministic, probabilistic, statistical, etc.

For the analysis levels, the tradeoff between the efficiency and the accuracy needs to be considered. The models with a higher level provide a design space with more degree of freedom for the reduction of power consumption [15]. As a result, a high-level model should be employed at the early design stage for its larger potential for power saving. On the other hand, due to its lack of design details, the accuracy of power estimation from high-level models is limited. Therefore, a good power management strategy should be: (1) Perform the power analysis at high level first in order to reduce the power consumption aggressively for the low-power design; (2) Perform low-level analysis next in order to enable further power reduction with high accuracy.

The choice of analysis engines in power analysis is again mainly determined by the tradeoffs between the simulation speed and the simulation accuracy. The SPICE-like simulators, although relatively slow in comparison with other general-purpose circuit simulators, offer the highest accuracy by directly solving the circuit nodal equations at the transistor level. Since there is no model simplification or approximation, exact transient simulation is done and thus the timing is well monitored. The importance of the timing information to power analysis is twofold. Firstly, the switching power is proportional to the running frequency (timing) of the chip. Secondly, the *toggle power* (due to the transient behaviors of signals before they are stable) are transient in nature, therefore the signal timing needs to be closely captured.

A novel power estimation method using SPICE as the analysis engine is the *power meter* technique [16], where the transient current drawn from the power supply is obtained by adding extra circuit elements to the original netlist. In order to improve computational efficiency, several other analysis engines such as fast timing simulator [17] and gate-level (logic) simulator [18] were used in power analysis. Because those simulators use groups of transistors as the basic simulation units, approximations must be made and certain loss of accuracy cannot be avoided.

The development of the power analysis techniques for improving the accuracy and efficiency of power estimation is still an important research area. In the rest of this chapter, the three distinct techniques, i.e., deterministic, probabilistic, and statistical techniques, will be described and discussed. The focus of the discussion will be on the statistical technique because of its accuracy, efficiency and simplicity.

2.4. INTRODUCTION TO POWER ANALYSIS TECHNIQUES

2.4.1 DETERMINISTIC POWER ANALYSIS

The *deterministic* technique, being strongly input pattern dependent, takes the user-specified primary input vectors and performs analysis at the specific level with the preferred analysis engine. This technique is clearly accurate because the inputs are known *a priori*. How the input vectors are collected and how many input sequences are needed to be representative are beyond the concerns of the deterministic technique. However, the input patterns can be generated exhaustively for all combination of possible input logic transitions when the total number of inputs is small, or are gathered for specific applications that are described by a sequence of architectural instructions. For the latter, it is often of great interest to provide the instructions that cause the maximum power (worst case). To generate such instructions in a processor design environment, for instance, the following constraints need to be considered:

- Number of instructions that can be dispatched per cycle
- Availability of various buffers and queues
- Execution time of various instructions
- Post dispatch serialization

The deterministic technique is exact as long as the design details are available. Unfortunately, it may not always be the case. Often the power analysis needs to be done for one block in early design phase before any other blocks in a chip are defined or completely specified. Therefore, usually the exact input specification is not available. In addition, when a chip needs to be qualified with its power rating, one may want to ensure that certain requirements are always met irrespective of the applications. For those cases, weakly input pattern dependent techniques, such as the probabilistic and statistical techniques, are preferred.

2.4.2 PROBABILISTIC POWER ANALYSIS

The *probabilistic* power analysis technique, simply put, is a way of calculating power by propagating the probability of logic transitions starting at the primary inputs. Because there are no exact input patterns required (i.e., only probabilities are required), it requires only one simulation run. Since no repeated simulation runs are necessary, it is computationally very efficient. One common limitation of the probabilistic techniques, however, is that they all require special delay models. The special delay models prohibit the use of existing simulation tools and libraries. Moreover, those models can significantly hinder the simulation accuracy.

To describe the concept of probabilistic techniques, we start with the following definitions:

Spatial Independence. Signals at the primary inputs or internal nodes in a circuit may be correlated. For instance, they may never be simultaneously high or low due to the logic topology. If the signal correlation is ignored in simulation, we call that the spatial independence is assumed.

Temporal Independence. Signal x at clock cycle T may be correlated to signal x at clock cycle $(T+1)$. For instance, the oscillator circuits expect to change state for every clock cycle. If the temporal correlation is ignored in simulation, we call that the temporal independence is assumed.

Signal Probability. The signal probability $P_s(x)$ at node x is the average fraction of clock cycles in which the stable logic value of x is high.

Transition Probability. The transition probability $\alpha_{a\to b}(x)$ at node x is the average fraction of clock cycles in which the logic value of x transitions from a to b. For instance, $\alpha_{0\to 1}$ stands for the probability of the logic transition from 0 to 1. Formally, $\alpha_{0\to 1}$ is defined as

$$\alpha_{0\to 1} = \lim_{N\to\infty} \frac{n(N)}{N} \tag{2.9}$$

where $n(N)$ is the number of $0 \rightarrow 1$ transitions in N clock cycles.

The probabilistic power analysis technique was first proposed in [19]. Both spatial and temporal independence was assumed in this original work. Consider a logic AND gate with the following boolean expression: $z = (a \cdot b)$ where $\cdot$ represents the AND operation. From the basic probability theory, if signals at input a and b are spatially independent, then $P_s(z) = P_s(a) \cdot P_s(b)$, where $P_s(\cdot)$ is the signal probability defined above. Similarly for a logic OR operation with the boolean expression $z = (a + b)$, the signal probability of z is

$$\begin{aligned} P_s(z) &= [1 - (1 - P_s(a)) \cdot (1 - P_s(b))] \\ &= P_s(a) + P_s(b) - P_s(a) \cdot P_s(b). \end{aligned} \tag{2.10}$$

To calculate the average power consumption in a circuit, the following formula can be used:

$$P_{avg} = f_{clk} \cdot V_{DD}^2 \sum_{i=1}^{n} C_L(x_i) \cdot \alpha_{0 \rightarrow 1}(x_i) \tag{2.11}$$

where f_{clk} is the clock frequency, $C_L(x_i)$ is the load capacitance at node x_i, and n is the total number of output nodes (of the logic gates) in the circuit. It is assumed that the dynamic switching power is dominant when using Eq. (2.11) to estimate the total power. Given the temporal independence assumption, $\alpha_{0 \rightarrow 1}$ can be computed in the following way. Consider a static 2-input NOR gate with the Boolean expression $z = \overline{a + b}$. Its transition probability is given by

$$\begin{aligned} \alpha_{0 \rightarrow 1}(z) &= P_s(z) \cdot P_s(\overline{z}) \\ &= P_s(z) \cdot [1 - P_s(z)] \end{aligned} \tag{2.12}$$

where

$$P_s(z) = 1 - P_s(a) - P_s(b) + P_s(a) \cdot P_s(b). \tag{2.13}$$

Assuming $P_s(a) = P_s(b) = 0.5$, the transition probability at the output of the NOR gate is 3/16.

Note that the transition probability $\alpha_{0 \rightarrow 1}$ depends on the *logic styles* (e.g., static logic, dynamic logic, etc). Equations (2.12) and (2.13) are valid only for the *static* NOR gate. For the dynamic logic, power is only dissipated when the output is switching from 1 to 0 during evaluation. As a result, the transition probability at the output of the dynamic NOR gate is 3/4.

The above analysis technique propagates the probability values from the primary inputs forward to the primary outputs. It is extremely fast because it takes advantage of the assumption of the signal spatial independence. Unfortunately, this assumption is rarely valid in real circuits, where the *reconvergent*

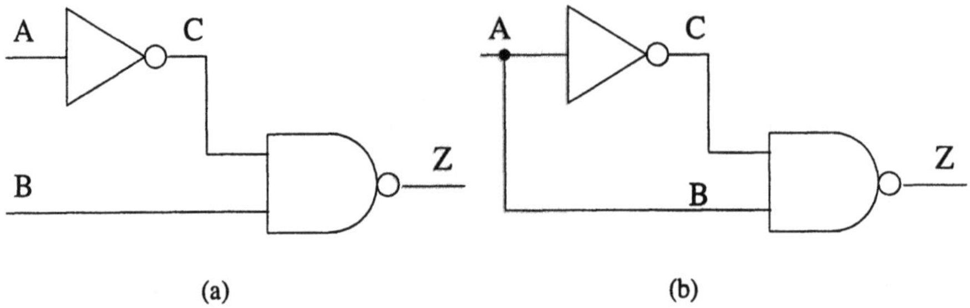

Figure 2.7. (a) Logic circuit without reconvergent fan-out, and (b) Logic circuit with reconvergent fan-out.

fan-out exists. Figure 2.7 illustrates this situation. In Fig. 2.7(a), signals at nodes C and B are independent, therefore the above probabilistic technique can be directly applied to find the signal and transition probabilities at node Z. In Fig. 2.7(b), however, the reconvergent fan-out exists. Signals at nodes B and C are inter-dependent because they all depend on the signal at node A. To analyze such a circuit, the described probabilistic technique needs to be extended by considering the *conditional probabilities*. Since the signal interdependence substantially complicates the correlation between signals, CAD tools are required for such analysis.

After the advent of the first probability-based power analysis tool, many other probabilistic approaches have been developed that aimed to improve the simulation accuracy. In [20], *probability waveforms* are specified at the primary input instead of fixed probability values. A probability waveform indicates the time period at which the signal logic is high, and the probability of signal transition from low to high at specific time points. With probability waveforms, the assumption of temporal independence is removed, and hence the accuracy is improved. Similarly, the equilibrium probability and transition density are specified at the primary inputs in [21] to eliminate the temporal independence assumption. Both techniques in [20] and [21] still assume the spatial independence among signals. The technique proposed in [22], which is based on the Binary Decision Diagram (BDD), attempts to handle both spatial and temporal correlations without the independence assumptions. As a result, it is very accurate compared to the previously developed probabilistic techniques. However, since BDD grows very rapidly and may even break down with increasing circuit size, its usefulness is mainly limited to moderate sized circuits.

2.4.3 STATISTICAL POWER ANALYSIS

The *statistical* technique for power analysis is an attractive choice among others because of its efficiency, accuracy, and simplicity. Its efficiency, although not as good as the probabilistic techniques, makes simulation of big circuits possible. It can be easily implemented in existing simulators.

The idea behind the statistical technique is to repeatedly simulate the circuit while monitoring the power being consumed. To proceed with our discussion, the following definition and statistical law are given first:

Sample Mean. Let $\mathbf{x}_1, \ldots, \mathbf{x}_n$ be a random sample of size n from some distribution with mean μ and variance σ^2. The *sample mean*, denoted $\overline{\mathbf{x}}_n$, is defined as the random variable obtained as the following arithmetic average:

$$\overline{\mathbf{x}}_n \triangleq \frac{\mathbf{x}_1 + \mathbf{x}_2 + \cdots + \mathbf{x}_n}{n} \tag{2.14}$$

Law of Large Numbers. As the number of random sample n increases, the distribution of the sample mean $\overline{\mathbf{x}}_n$ gets tightened around its distribution mean μ. In the limit, it can be expressed as

$$\lim_{n \to \infty} \overline{\mathbf{x}}_n = \mu. \tag{2.15}$$

The *Law of Large Numbers* is the fundamental principle behind the statistical technique. It illustrates how the repeated simulation can estimate the average value (i.e., mean) of a random variable, as long as the number of samples is large enough.

In order to obtain the statistical measure of the transition activities (power) in a circuit, the statistical characteristics of the primary inputs must be specified. An *input pattern generator* is then used to generate the input vectors for the statistical simulation based on the given input characteristics. The next outstanding question is, how many input vectors are enough. Or, how many random samples do we need *at least* so that the Law of Large Numbers holds. A statistical power analysis tool addresses the above question.

A standard statistical power estimation flow is shown in Fig. 2.8. Detailed implementation of each component in this flow varies among different analysis tools. For the circuit simulation part, one can choose from any of the analysis engines described earlier in this chapter. For the input pattern generation part, if the statistical characteristics of the input is unknown at the time of power estimation or the input stream does not contain signal correlations, it is usually sufficient to use a random number generator. Alternatively, the input vectors can be directly drawn from the input stream pool if provided.

The statistical measurement block in Fig. 2.8 is needed in order to collect and update the statistical data. Next, the stopping criterion is used to determine

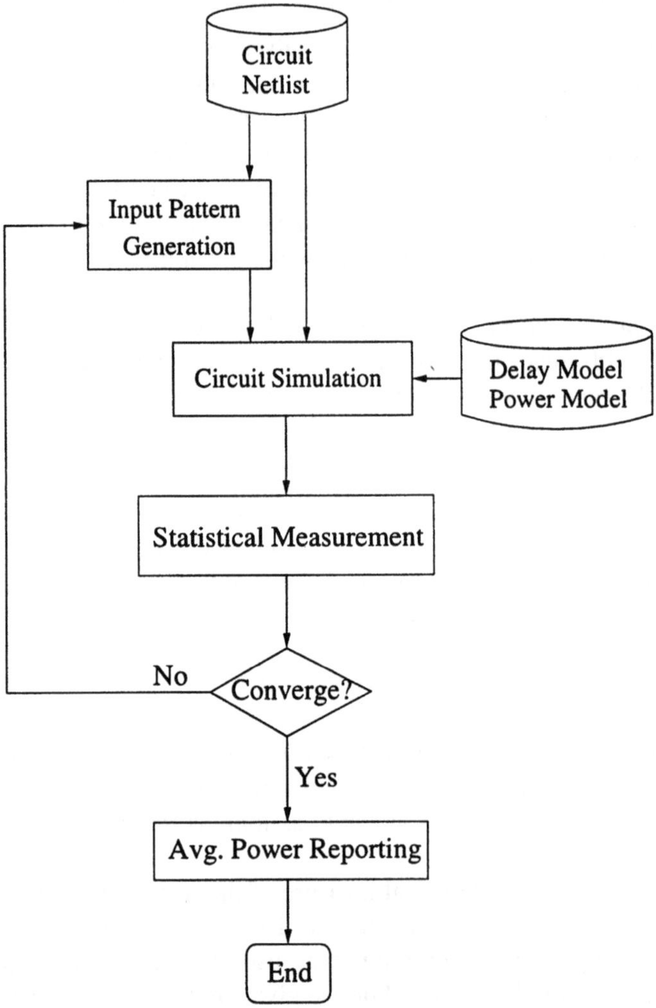

Figure 2.8. A standard statistical power estimation flow.

whether the average power estimated thus far is close enough (converged) to the real value. If it is, the repeated simulation is terminated and the average power value is reported. Otherwise, another input pattern will be generated for the next round of circuit simulation. For simulation efficiency, it is crucial that the stopping criterion will result in the sample size as small as possible, while the required accuracy is achieved.

To derive a stopping criterion, the *central limit theorem* and the concept of *confidence level* are often used [23]:

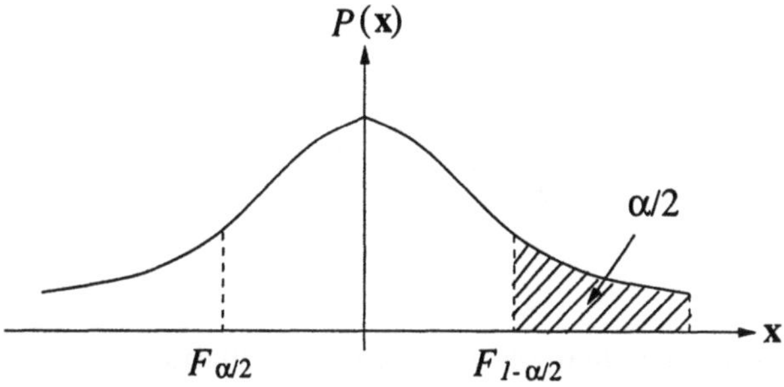

Figure 2.9. Relationship between F, α, and the confidence level.

Central Limit Theorem. If $\mathbf{x}_1, \ldots, \mathbf{x}_n$ form a random sample from an *arbitrary* distribution, with mean μ and variance σ^2, and if $\overline{\mathbf{x}}_n$ is their sample mean, then

$$\lim_{n \to \infty} P\left\{\frac{\overline{\mathbf{x}}_n - \mu}{\sigma/\sqrt{n}} \leq x\right\} = \Phi(x) \tag{2.16}$$

where P is the probability and $\Phi(x)$ is the *cumulative distribution function* (cdf) of the standard normal distribution. In other words, the random variable $\frac{\overline{\mathbf{x}}_n - \mu}{\sigma/\sqrt{n}}$ has the standard normal distribution for large n.

Confidence Level. If a random variable $\mathbf{x}$ has cdf $F(\cdot)$, and $0 \leq \alpha \leq 1$, then the α-percentile of $F(\cdot)$, denoted F_α, is a real number such that:

$$P\{\mathbf{x} \leq F_\alpha\} = \alpha. \tag{2.17}$$

It then follows that

$$P\{\mathbf{x} \leq F_{\alpha/2}\} = \alpha/2 \tag{2.18}$$
$$P\{\mathbf{x} \leq F_{1-\alpha/2}\} = 1 - \alpha/2 \tag{2.19}$$

and

$$P\{F_{\alpha/2} < \mathbf{x} \leq F_{1-\alpha/2}\} = (1 - \alpha/2) - (\alpha/2) = 1 - \alpha. \tag{2.20}$$

The above equation can be interpreted as follows: With $(1 - \alpha)$ confidence, the inequality $F_{\alpha/2} < \mathbf{x} \leq F_{1-\alpha/2}$ is valid (See also Fig. 2.9).

Given that $\frac{\overline{\mathbf{x}}_n - \mu}{\sigma/\sqrt{n}}$ has the standard normal distribution for large n (central limit theorem) and from Eq. (2.20), it follows that

$$P\{\Phi_{\alpha/2} < \frac{\overline{\mathbf{x}}_n - \mu}{\sigma/\sqrt{n}} \leq \Phi_{1-\alpha/2}\} = 1 - \alpha. \tag{2.21}$$

Since $\Phi_{\alpha/2} = -\Phi_{1-\alpha/2}$ (See Fig. 2.9), Eq. (2.21) can be recast to

$$\frac{|\mu - \overline{\mathbf{x}}_n|}{\overline{\mathbf{x}}_n} \leq \frac{\Phi_{1-\alpha/2}\sigma}{\overline{\mathbf{x}}_n\sqrt{n}} \leq \epsilon \tag{2.22}$$

where ϵ is the user specified *error tolerance*. Equation (2.22) can be readily used as a stopping criterion for the statistical power estimation: Once users specify ϵ and the the confidence level $(1 - \alpha)$, the minimally required number of repeated simulations (i.e., n) can be calculated according to Eq. (2.20). Note that for the standard normal distribution, Φ is well tabulated and σ can be easily calculated.

McPower [24] was developed to statistically estimate the *total* average power of a circuit by using the Monte-Carlo simulation. It is assumed in this approach that the power consumed by a circuit over time has a distribution close to normal. Based on this assumption, a set of stopping criterion similar to Eq. (2.22) is derived for the given confidence level and percentage error. This approach assumes no spatial and temporal independence for internal nodes. It only requires that the primary inputs be independent and do not change with time in statistical characteristics.

Another statistical technique, called MED [25], was developed for estimating the average power of *each gate* in a circuit. It is an attractive technique because it provides a very useful piece of information: *power density* i.e., power consumption per unit area. This localized information is essential for electrothermal simulation to find the hot spots caused by the gates (or macros) with high power density. The MED approach is conceptually similar to McPower, except for its stopping criteria. Two stopping criteria are needed in MED for good convergence rate, one for the normal gates and the other for gates with very few logic transitions. The details of the MED approach and its application to the electrothermal simulation environment will be presented in Chapter 8 for the temperature-driven power and timing analysis.

The McPower-like approaches assume that the average power follows the normal distribution based on the central limit theorem. Such approaches are called the *parametric* approaches because they are designed by employing the properties of certain statistical distribution function. Another approach, called the *nonparametric* approach, was proposed in [26] for estimating the average power of a circuit. It is a reliable alternative for average power analysis because no up-front statistical distribution function of the average power is assumed. It was shown that the nonparametric approach can generate more accurate and robust average estimates with higher simulation costs.

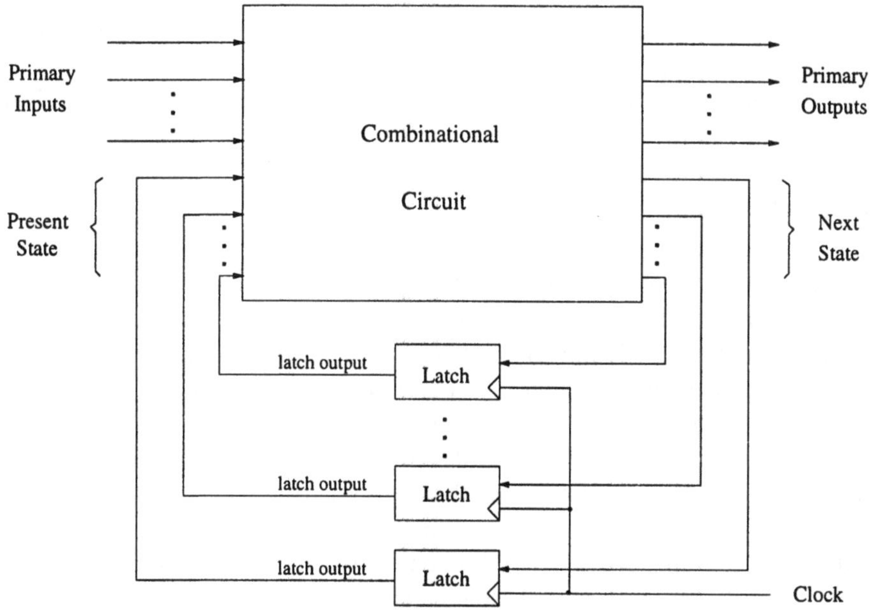

Figure 2.10. A generic sequential circuit.

2.4.4 POWER ANALYSIS FOR SEQUENTIAL CIRCUITS

All of the above power analysis techniques either assume certain input statistics, or at minimum assume that the inputs are independent. Those are reasonable assumptions for combinational circuits because unless exact input vectors are specified as in the deterministic approach, the circuit's input patterns are unknown by nature. For sequential circuits, however, none of the above approaches can be directly applied.

In order to implement a finite-state machine (FSM), a sequential circuit contains latches and feedback loops, thus introducing spatio-temporal correlations among the latch output signals. It is shown in Fig. 2.10, where the present state lines form the secondary inputs to the combinational circuit. Because the present state lines are the outputs of latches, they are not necessarily independent and must be characterized.

Power and switching activity estimation for sequential circuits are significantly more difficult than for combinational circuits because the number of states in an FSM increases exponentially with the number of latches. To reduce the complexity of power analysis for sequential circuits to a manageable level, it is convenient to partition the circuits and perform power analysis for the latches and the combinational blocks separately. Most existing probabilistic or statistical power analysis techniques for combinational circuits introduced earlier can be directly utilized, and the only extra information needed is the transition probability or statistics of the latch outputs (secondary inputs). This information can be gathered by either solving the state equations or by logic simulation. In the following, a brief overview of some distinct techniques is presented.

In [27], an exact probabilistic method for estimating the sequential circuit power with the given *state transition graph* (STG) of the FSM, is presented. The state probabilities (i.e., probability for an FSM being in a particular state) are first calculated by iteratively solving the Chapman-Kolmogorov equations [28]. Using these probabilities, the exact signal probabilities (i.e., probability of a signal being logic high) and transition probabilities (i.e., probability of a signal that will switch) are calculated by an implicit state enumeration procedure. The average power can then be directly found from the transition probabilities. Although this exact method is computational very expensive and thus its usefulness is severely limited by the circuit size, many subsequent approximation approaches are based on its formulation.

In [27], an approximation method is presented, which solves the nonlinear system equations to find the present state line probabilities. This nonlinear system is represented by $P_{ns} = F(P_{ps})$, where P_{ps} and P_{ns} are vectors of the state line probabilities for present state and next state, respectively. The function $F(\cdot)$ is a nonlinear function determined by the Boolean equation of the combinational logic shown in Fig. 2.10; $F(\cdot)$ can be obtained from a BDD. From Fig. 2.10 it can be seen that the present state lines are the outputs of latches, therefore $P_{ps} = P_{ns} \equiv P$, and the nonlinear equation can be reduced to $P = F(P)$. This system is then solved by applying the Picard-Peano iteration method. In order to correct the assumption that the presented state lines are independent, the authors in [27] proposed a method to unroll and cascade the combinational logic k times such that the effect of the independence assumption becomes mild.

A totally new approach for power analysis in sequential circuits is described in [29], which is based on a statistical estimation technique. By applying randomly generated input sequences to the circuit, statistics on the latch outputs are collected by logic simulation. Like statistical techniques developed for the combinational circuits, a set of stopping criteria determining how many random samples are needed and how long (in terms of clock cycles) the FSM needs to be simulated for each sample, are derived for the sequential circuits. After the statistics of the latch outputs are collected, it is then possible to use any of the existing combinational circuit techniques to compute the total circuit power. One important advantage of this approach is that the desired accuracy of the simulation results can be specified upfront by users.

In [30], another statistical technique was proposed for sequential circuit power estimation. An *initial transient problem* was investigated and a solution was provided. In an FSM, a set of states is said to be *closed* if no state outside can be reached from any state in it. If such a closed set does exist and if the randomly generated initial state of the FSM is in this set, the estimated power can be severely biased. In other words, the estimated power with the initial state in closed set $S1$ may be very different from the one with he initial state in closed set $S2$. The above is called the initial transient problem.

A concept of *warm-up period* is given in [30] to address the initial transient problem. In theory, the state of an FSM at time k becomes independent of its initial state at time 0 as $k \to \infty$, assuming that all logic signals are non-periodic. The warm-up period is the period of time that the FSM has to be simulated before its power values is independent of the initial state. Therefore, the power data should be collected only *after* the warm-up period for each sample.

In [31], a nonparametric statistical approach for sequential circuit power analysis is presented. The concept of warm-up period is also visited and the length of warm-up period is determined by the *hypothesis test* for randomness. A hypothesis test determines whether or not a power sequence is a random sample by the statistical measure of that sequence. Given a specified *significance level*, if the hypothesis is *accepted* at a trial independence interval, the sequence can be viewed as a random sample. Otherwise the trial interval is increased to reduce the temporal correlation. These steps are repeated in an iterative manner until the randomness hypothesis is accepted and the selected interval is used hereafter to generate random power samples. Next, the stopping criterion based on the order-statistics is used to analyze the sample data and control the sample size until the desired accuracy is achieved. In addition to high accuracy by considering all signal correlations, the simulation efficiency is also greatly improved by choosing the independence interval dynamically.

2.5. SUMMARY

This chapter provides an overview of various power analysis techniques for CMOS circuits. The estimated power values can be used by thermal analysis to calculate the temperature distribution.

- Sources of power consumption in CMOS technology are described:
  - Dynamic (switching) power
  - Internal power
  - Short-circuit power
  - Leakage power
- Three different criteria based on which a power analysis tool can be classified are introduced:

- Analysis level
- Analysis engine
- Analysis technique

- The deterministic power analysis technique is first presented. A number of constraints that need to be considered when generating a good set of input vectors are given. The limitation of this technique due to its input pattern dependence is described.

- The probabilistic power analysis technique is presented. Some assumptions that are commonly used in this technique are provided. The definition of signal probability and transition probability is also given. Examples of calculating the signal probability and transition probability for simple logic gates are provided. Reconvergent fan-outs in a circuit is considered for finding the (conditional) probabilities. Finally, different probabilistic power analysis tools are compared.

- The statistical power analysis technique is presented. Some commonly used statistical terms and theorems are provided. A standard statistical power analysis flow is shown. Based on the central limit theorem, the stopping (convergence) criterion is derived so that the repeated simulation is terminated while the specified accuracy level is reached. Two branches of statistical power analysis techniques are also described: parametric and nonparametric techniques. Different statistical power analysis tools are presented.

- The above power analysis techniques are revisited for analyzing the sequential circuits. Because the sequential circuit contains memory elements and feedbacks, the inputs to the circuits can no longer be assumed independent. A number of approaches used to take into account such spatio-temporal correlations are illustrated.

References

[1] N. R. Shanbhag, "Lower bounds on power-dissipation for dsp algorithms," in *Proceedings of the International Symposium on Low Power Electronics and Design*, pp. 43–48, 1996.

[2] A. Tyagi, "Entropic bounds on fsm switching," in *Proceedings of the International Symposium on Low Power Electronics and Design*, pp. 323–328, 1996.

[3] S. Ramprasad, N. R. Shanbhag, and I. N. Hajj, "Achievable bounds on signal transition activity," in *Proceedings of the ACM/IEEE International Conference on Computer-Aided Design*, pp. 126–129, 1997.

[4] D. Marculescu, R. Marculescu, and M. Pedram, "Theoretical bounds for switching activity analysis in finite-state machines," in *Proceedings of the International Symposium on Low Power Electronics and Design*, pp. 36–41, 1998.

[5] L. P. Yuan, *Power and voltage drop analyses in VLSI circuits.* PhD thesis, Dept. of Electrical and Computer Engineering, University of Illinois at Urbana-Champaign, 1999.

[6] C. Y. Tsui, M. Pedram, and A. M. Despain, "Power efficient technology decomposition and mapping under an extended power consumption model," *IEEE Transactions on Computer-Aided Design of Integrated Circuits and Systems*, vol. 13, no. 9, pp. 1110–1122, 1985.

[7] H. J. M. Veendrick, "Short-circuit dissipation of static CMOS circuitry and its impact on the design of buffer circuits," *IEEE Journal of Solid-State Circuits*, vol. 19, no. 8, pp. 468–473, 1984.

[8] N. Hedenstierna and K. O. Jeppson, "CMOS circuit speed and buffer optimization," *IEEE Transactions on Computer-Aided Design of Integrated Circuits and Systems*, vol. 6, no. 2, pp. 270–281, 1987.

[9] T. Sakurai and A. R. Newton, "Alpha-power law MOSFET model and its applications to CMOS inverter delay and other formulas," *IEEE Journal of Solid-State Circuits*, vol. 25, no. 2, pp. 584–594, 1990.

[10] S. R. Vemuru and N. Scheinberg, "Short-circuit power dissipation estimation for CMOS logic gates," *IEEE Transactions on Circuits and Systems*, vol. 41, no. 11, pp. 762–765, 1994.

[11] A. Alvandpour, P. Larsson-Edefors, and C. Svensson, "Separation and extraction of short-circuit power consumption in digital CMOS VLSI circuits," in *Proceedings of the International Symposium on Low Power Electronics and Design*, pp. 245–258, 1998.

[12] A. Chandrakasan, I. Yang, C. Vieri, and D. Antoniadis, "Design consideration and tools for low-voltage digital system design," in *Proceedings of the ACM/IEEE Design Automation Conference*, pp. 113–118, 1996.

[13] Z. Chen, M. Johnson, L. Wei, and K. Roy, "Estimation of standby leakage power in CMOS circuits considering accurate modeling of transistor

stacks," in *Proceedings of the International Symposium on Low Power Electronics and Design*, pp. 239–244, 1998.

[14] J. P. Halter and F. Najm, "A gate-level leakage power reduction method for ultra-low-power CMOS circuits," in *Proceedings of the IEEE Custom Integrated Circuits Conference*, pp. 475–478, 1997.

[15] M. Nemani and F. Najm, "High-level area and power estimation for VLSI circuits," *IEEE Transactions on Computer-Aided Design of Integrated Circuits and Systems*, vol. 18, no. 6, pp. 697–713, 1999.

[16] S. M. Kang, "Accurate simulation of power dissipation in VLSI circuits," *IEEE Journal of Solid-State Circuits*, vol. 21, pp. 889–891, Oct. 1986.

[17] Y. K. Cheng, P. Raha., C. C. Teng, E. Rosenbaum, and S. M. Kang, "ILLIADS-T: An electrothermal timing simulator for temperature-sensitive reliability diagnosis of CMOS VLSI chips," *IEEE Transactions on Computer-Aided Design of Integrated Circuits and Systems*, vol. 17, pp. 668–681, Aug. 1998.

[18] F. Dresig, P. Lanches, O. Rettig, and U. G. Baitinger, "Simulation and reduction of CMOS power dissipation at logic level," in *Proceedings of the European Design Automation Conference*, pp. 341–346, 1993.

[19] M. A. Cirit, "Estimating dynamic power consumption of CMOS circuits," in *Proceedings of the ACM/IEEE International Conference on Computer-Aided Design*, pp. 534–537, 1987.

[20] F. Najm, R. Burch, P. Yang, and I. Hajj, "CREST: A current estimator for CMOS circuits," in *Proceedings of the ACM/IEEE International Conference on Computer-Aided Design*, pp. 204–207, 1988.

[21] F. Najm, "Transition density: a new measure of activity in digital circuits," *IEEE Transactions on Computer-Aided Design of Integrated Circuits and Systems*, vol. 12, no. 2, pp. 310–323, 1993.

[22] R. E. Bryant, "Graph-based algorithms for Boolean function manipulation," *IEEE Transactions on Computers*, vol. 35, pp. 677–691, Aug. 1986.

[23] P. L. Meyer, *Introductory Probability and Statistical Applications*. Addison-Wesley, 1970.

[24] R. Burch, F. Najm, P. Yang, and T. Trick, "McPOWER: A Monte Carlo approach to power estimation," in *Proceedings of the ACM/IEEE International Conference on Computer-Aided Design*, pp. 90–97, 1992.

[25] M. G. Xakellis and F. N. Najm, "Statistical estimation of the switching activity in digital circuits," in *Proceedings of the ACM/IEEE Design Automation Conference*, pp. 728–733, June 1994.

[26] L. P. Yuan, C. C. Teng, and S. M. Kang, "Nonparametric estimation of average power dissipation in CMOS VLSI circuits," in *Proceedings of the IEEE Custom Integrated Circuits Conference*, pp. 225–228, 1996.

[27] C. Y. Tsui, M. Pedram, and A. M. Despain, "Exact and approximate methods for calculating signal and transition probabilities in FSMs," in *Proceedings of the ACM/IEEE Design Automation Conference*, pp. 18–23, 1994.

[28] A. Papoulis, *Probability, Random Variables and Stochastic Processes.* New York: McGraw-Hill, 1984.

[29] F. Najm, S. Goel, and I. Hajj, "Power estimation in sequential circuits," in *Proceedings of the ACM/IEEE Design Automation Conference*, pp. 635–640, 1995.

[30] T. L. Chou and K. Roy, "Statistical estimation of sequential circuit activity," in *Proceedings of the ACM/IEEE International Conference on Computer-Aided Design*, pp. 34–37, 1995.

[31] L. P. Yuan, C. C. Teng, and S. M. Kang, "Statistical estimation of average power dissipation in sequential circuits," in *Proceedings of the ACM/IEEE Design Automation Conference*, pp. 377–382, 1997.

Chapter 3

TEMPERATURE-DEPENDENT MOS DEVICE MODELING

3.1. INTRODUCTION

A circuit can be represented by a set of node equations describing the circuit network comprised of semiconductor devices. To accurately model the thermal effect in a circuit, the device model must take into account the device temperature as one of its modeling parameters.

Due to the rapid advance in semiconductor fabrication technologies, the device feature size has been drastically reduced to the very deep submicron regime. The short channel effects in devices often necessitate the use of 3D device modeling or sometimes quantum physics for device analysis. One of the device simulation techniques, namely the *Monte-Carlo* simulation, investigates the electron scattering mechanisms in the crystal from the probability view point. With the increasing speed and capacity of computers today, one can accurately model the semiconductor device characteristics, such as electronic current, using this technique.

Monte-Carlo simulation for semiconductor devices requires solving the Boltzmann equation coupled with the Poisson equation. It is often very complicated, and assumptions are made so that simple forms of modeling equations can be derived. Once the simplified device equations and models are found, their physical or fitting parameters can be determined on the basis of physical properties of the devices, measured data, or combination of the two. They can then be readily applied to the large-scale circuit simulation and analysis with sufficient accuracy.

In the following, an overview of the temperature-dependent device physics and device modeling will be presented. The temperature-dependent models of the threshold voltage and carrier mobility of the MOS transistors are derived. Two device models that are suitable for the temperature-sensitive circuit simulation will be visited as case studies.

3.2. TEMPERATURE-DEPENDENT DEVICE PHYSICS AND MODELING

The popular Shichman-Hodges model [1] (also known as the SPICE Level 1 model) for the MOS transistor defines the MOS current in two operating regions in terms of the gate-source voltage V_{GS} and the drain-source voltage V_{DS} as follows:

$$I_{DS} = \begin{cases} \beta(2V_{GSE}V_{DS} - V_{DS}^2) & \text{if } V_{DS} \leq V_{GSE} \text{ (linear)} \\ \beta V_{GSE}^2 & \text{if } V_{DS} > V_{GSE} \text{ (saturation).} \end{cases} \quad (3.1)$$

In the above equation, β is the MOS device transconductance given by $\frac{1}{2}\mu_0 C_{ox}\frac{W}{L}$ and $V_{GSE} = V_{GS} - V_T$, where μ_0 is the carrier mobility, C_{ox} is the gate oxide capacitance per unit area, V_T is the threshold voltage, W and L are the device channel width and length respectively. The channel-length modulation parameter λ is ignored in Eq. (3.1) for simplicity. Equation (3.1) contains two temperature-dependent physical parameters, μ_0 and V_T.

3.2.1 TEMPERATURE-DEPENDENT THRESHOLD VOLTAGE

The threshold voltage V_T can be expressed as the summation of the zero-bias threshold voltage V_{T0} and another term related to the body effect. The temperature dependence in V_{T0} is contributed by several material-related physical parameters and can be derived by applying MOS device physics. In the SPICE Level 1 MOS model, for instance, $V_{T0}(T)$ of an NMOSFET with n^+-polysilicon gate at temperature T can be formulated as

$$V_{T0}^n(T) = -\frac{E_g}{2q} + \frac{kT}{q}\ln\frac{N_a}{n_i} - \frac{Q_{tot}}{C_{ox}} + \frac{2}{C_{ox}}\sqrt{\epsilon_{si}kT(\ln\frac{N_a}{n_i})N_a}. \quad (3.2)$$

Similarly, $V_{T0}(T)$ of a PMOSFET with n^+-polysilicon gate is

$$V_{T0}^p(T) = -\frac{E_g}{2q} - \frac{kT}{q}\ln\frac{N_d}{n_i} - \frac{Q_{tot}}{C_{ox}} - \frac{2}{C_{ox}}\sqrt{\epsilon_{si}kT(\ln\frac{N_d}{n_i})N_d} \quad (3.3)$$

where

$$n_i = 3.9 \times 10^{16} T^{1.5} e^{-(E_g/2kT)}, \text{ and} \quad (3.4)$$
$$E_g = 1.179 - 9.025 \times 10^{-5}T - 3.05 \times 10^{-7}T^2. \quad (3.5)$$

In the above equations, n_i is the intrinsic carrier concentration of silicon, E_g is the energy band gap of silicon, N_a and N_d are the doping concentration,

k is the Boltzmann constant, and ϵ_{si} is the permittivity of silicon. The total charge density Q_{tot} includes the surface state charge density (Q_{ss}), fixed charge density (Q_F), and the threshold voltage-adjustment implant density (Q_{imp}).

3.2.2 TEMPERATURE-DEPENDENT CARRIER MOBILITY

The carrier mobility in semiconductors is related directly to the mean-free time between collisions, which in turn is determined by various scattering mechanisms. The three most important mechanisms are Coulomb, lattice, and surface-roughness scatterings.

COULOMB (IMPURITY) SCATTERING. Coulomb scattering results when a charge carrier travels past an ionized impurity. The effect of Coulomb scattering at high temperature is small because the carriers are moving faster and, therefore, scatter less. However, this cannot be neglected because the oxide charges contribute to the Coulomb scattering at room temperature or higher [2]. In [2], it is proven that the Coulomb-scattering-limited mobility μ_C follows

$$\mu_C \propto T/N_I \tag{3.6}$$

where T is temperature and N_I is the charge density at the Si-SiO_2 interface.

LATTICE (PHONON) SCATTERING. Lattice scattering results from thermal vibrations of the lattice atoms at any temperature above zero. These vibrations disturb lattice periodic potential and allow energy to be transferred between the carriers and the lattice. For intermediate inversion-layer concentrations ($Q_N/q = 0.5 \sim 5 \times 10^{12}/\text{cm}^2$), the channel mobility has been observed to have the following relationships with the effective transverse electric field E_{eff} and T [3]:

$$\mu_L \propto T^{-n} E_{eff}^{-1/\gamma} \tag{3.7}$$

where

$$E_{eff} = \frac{(0.5 \cdot Q_N + Q_D)}{\epsilon_{si}}. \tag{3.8}$$

The Q_D term is the depletion charge density, $\gamma = 3 \sim 6$, and $n = 1 \sim 1.5$, depending on the crystallographic orientation and the strength of intervalley and intersubband scattering.

SURFACE-ROUGHNESS SCATTERING. Surface-roughness scattering results from the asperities at the Si-SiO_2 interface at high electron concentrations. The dependence of the surface-roughness scattering-limited mobility

μ_{SR} on E_{eff} is given by [4]

$$\mu_{SR} \propto E_{eff}^{-2}. \tag{3.9}$$

Because the probability of a collision ($1/\tau_c$) taking place in unit time is the sum of the probabilities of collisions due to various scattering mechanisms, it follows that

$$\frac{1}{\tau_c} = \frac{1}{\tau_{c,C}} + \frac{1}{\tau_{c,L}} + \frac{1}{\tau_{c,SR}}.$$

Or equivalently:

$$\frac{1}{\mu_n} = \frac{1}{\mu_C} + \frac{1}{\mu_L} + \frac{1}{\mu_{SR}}. \tag{3.10}$$

The simple temperature-dependent mobility formula used in the SPICE Level 1 model is

$$\mu_0(T) = \mu_0(300\text{K}) \times (\frac{T}{300\text{K}})^{-1.5} \tag{3.11}$$

where lattice scattering is assumed to be the only factor that determines the carrier mobility, and n in Eq. (3.7) is chosen to be 1.5.

3.3. TEMPERATURE-DEPENDENT BSIM MODEL FOR SPICE SIMULATION

In [5], an accurate temperature-dependent device modeling technique for MOS VLSI circuit simulation was proposed. This technique has been incorporated into the Berkeley Short-Channel IGFET Model (BSIM, also called the Level-4 model) to predict the temperature-dependent characteristics of devices in the submicron regime.

Each electrical parameter P_i in BSIM is described by the following formula [6]:

$$P_i = P_{0i} + \frac{P_{Li}}{L_{eff}} + \frac{P_{Wi}}{W_{eff}} \tag{3.12}$$

where P_{0i}, P_{Li} and P_{Wi} are the process parameters associated with the electrical parameter P_i. The effective channel length (L_{eff}) and channel width (W_{eff}) are used in the formula. The parameters P_{0i}, P_{Li} and P_{Wi} are generated by fitting the parameter file with different device dimensions to Eq. (3.12). Once P_{0i}, P_{Li} and P_{Wi} are known, the electrical parameters for a device with any channel length and width can be calculated from Eq. (3.12) by replacing L_{eff} and W_{eff} with the desired dimensions.

Because each electrical parameter is calculated from several process parameters, the BSIM contains sixty-seven process parameters in total. Furthermore, the BSIM drain current equation is considerably more complicated than the simple formula in Eq. (3.1). In the triode (linear) region, for instance, the BSIM drain current can be written as:

$$I_{DS} = \frac{\mu_0 C_{ox} \frac{W}{L}[(V_{GS} - V_T)V_{DS} - \frac{a}{2}V_{DS}^2]}{[1 + U_0(V_{GS} - V_T)](1 + \frac{U_1}{L}V_{DS})}. \tag{3.13}$$

In this equation, μ_0, C_{ox}, V_T, a, U_0, and U_1 can all be expressed in terms of the BSIM electrical parameters. Totally eighteen different electrical parameters are needed in calculating the drain current expressed in Eq. (3.13). If the temperature dependence needs to be considered, even more complexity will be added to Eq. (3.13).

One straightforward way to model the temperature-dependent BSIM drain current is to characterize each electrical parameter used in Eq. (3.13) over a certain range of temperatures. In other words, P_i becomes $P_i(T)$ in Eq. (3.12) and $I_{DS}(T)$ can be found accordingly by Eq. (3.13). Unfortunately, this approach is computationally expensive. One attractive alternative proposed by [5] utilizes the concept of *sensitivity*. It first determines a sensitive SPICE parameter subset that has large effects on the device output characteristics, and only the parameters in this subset need to be updated as temperature changes.

To find the sensitive subset, a sensitivity function is used. A basic sensitivity function of a variable Q with respect to an electrical parameter P_i is defined as [7]:

$$S_{Q|P_i} = (\frac{\partial Q}{\partial P_i})/(\frac{Q}{P_i}). \tag{3.14}$$

For digital ICs, the variable Q could be the drain current of the MOS transistor. A modified sensitivity function for the drain current over a biasing space A (e.g., linear, saturation) can be expressed as:

$$\hat{S}_{I_{DS}|P_i} = \frac{1}{(\partial P_i/P_i)}\sqrt{\frac{1}{m}\sum_{A}\left[\frac{I_{DS} - I_{DS,P_i}}{\max(I_{DS}, I_{DS0})}\right]^2} \tag{3.15}$$

where I_{DS,P_i} is the simulated drain current with P_i being perturbed and m is the number of data points. The user-specified input I_{DS0} is used to determine whether an absolute or relative error is preferred.

After the sensitivity analysis, there are *eight* out of eighteen BSIM electrical parameters left in the sensitivity subset. It means that if the BSIM drain current needs to be updated due to any environmental perturbation, including temperature change, only eight parameters need to be recalculated. Table 3.1 lists these eight electrical parameters, most of which are related to the carrier mobility and the threshold voltage. The above procedure for finding the sensitive parameter subset corresponds to the *sensitivity analysis* box in Fig. 3.1, which illustrates the whole BSIM sensitive parameter subset approach.

Table 3.1. BSIM sensitive parameters

Parameter Names	Physical Meaning
V_{FB}	Flat-band voltage
ϕ_s	Surface inversion potential
K_1	Body effect coefficient
K_2	Drain/source depletion charge sharing coefficient
μ_z	Zero-bias mobility
U_{0z}	Zero-bias transverse-field mobility degradation coefficient
U_{1z}	Zero-bias velocity saturation coefficient
μ_s	Mobility at zero substrate bias and at $V_{DS} = V_{DD}$

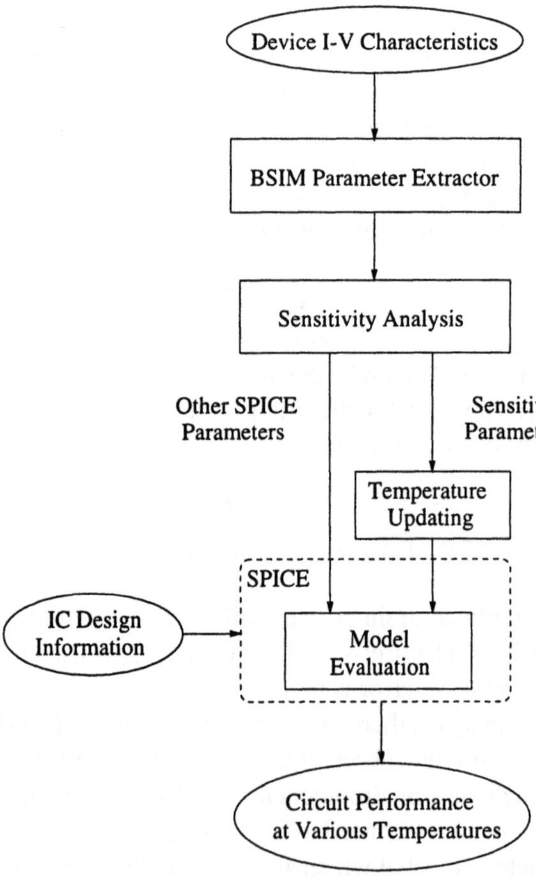

Figure 3.1. BSIM sensitive parameter subset approach.

Next, in order to determine how sensitive those eight parameters are to the temperature variation, their temperature coefficients ($\partial P_i/\partial T$) need to be found. The temperature coefficients can be obtained from one of the following two sources: the measured experimental data or the analytically derived formula. In the first case, the values of process parameters are extracted automatically from the device I-V data using the BSIM_Extract program at different temperatures. The parameter-temperature relationship can thus be found and fitted by an appropriate fitting function. For instance, values of the parameters K_1, K_2, U_{0z}, V_{FB}, and ϕ_s are fitted by a linear function of temperature, while values of the parameters μ_z and μ_s are fitted by functions of $T^{-1.5}$ and values of U_{1z} are fitted by $T^{-2.5}$ [5].

The temperature coefficients can also be analytically derived, as will be illustrated by taking ϕ_s as an example. The surface inversion potential (ϕ_s) is given by

$$\phi_s = \frac{2kT}{q}\log(\frac{N_a}{n_i}) \tag{3.16}$$

where n_i is also temperature dependent as shown in Eq. (3.4). Its temperature coefficient can be derived and represented by a simple analytical formula [8]:

$$\frac{\partial \phi_s}{\partial T} \simeq -\frac{1}{T}(\frac{E_g(T=0)}{q} - \phi_s). \tag{3.17}$$

By using Eq. (3.17), parameter ϕ_s can be easily updated under any temperature perturbation during the SPICE simulation. Consequently, the drain current can be recalculated efficiently. This can be done for all other sensitive parameters as long as simple yet accurate analytical formula can be obtained. Figure 3.2 demonstrates the procedure for incorporating the temperature coefficient values into the BSIM parameter update, where P_{0i}, P_{Li} and P_{Wi} are the process parameters used in Eq. (3.12).

3.4. REGIONWISE QUADRATIC (RWQ) MODEL

The fast-timing simulation technique was originally proposed to bridge the gap between full timing simulation and switch-level timing simulation. It is computationally very efficient compared to SPICE-like simulation method. The description of the general fast-timing simulation technique and the introduction of the fast timing simulator developed at the University of Illinois, named ILLIADS, will be given in Chapter 5. In this section, the device modeling technique employed by ILLIADS, called the regionwise-quadratic (RWQ) model [9], is discussed. The extension of the RWQ model to take into account the temperature dependence is also addressed in detail.

The RWQ model was developed for submicron device modeling in order to improve the accuracy of fast timing simulation. Its modeling procedure takes as input a set of data points (V_{DS}, V_{GSE}, I_{DS}) that have been obtained either from

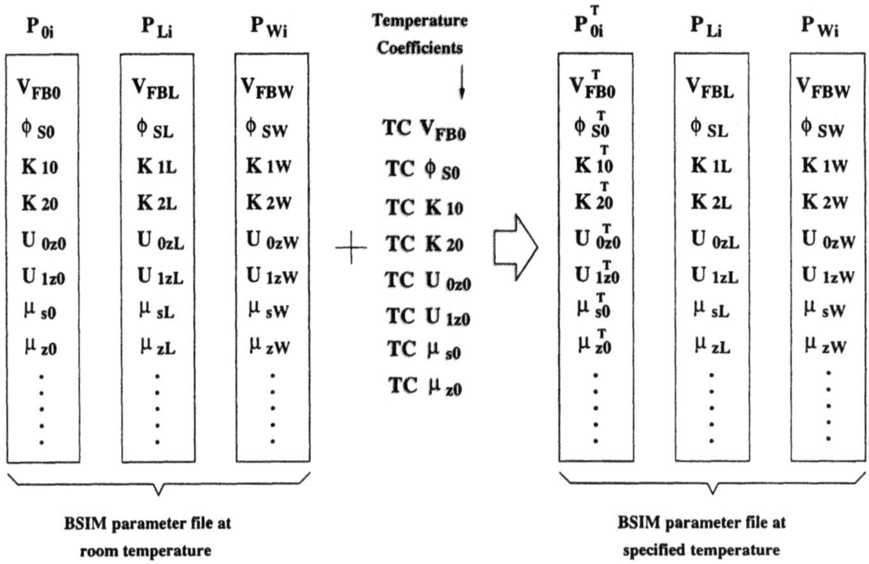

Figure 3.2. BSIM parameter value update using temperature coefficients.

measured data of a test device or by exercising (using SPICE, for example) a particular analytical or empirical MOS I-V model. Here $V_{GSE} = V_{GS} - V_{T0}$ and V_{T0} is the zero-bias threshold voltage of a MOS device. Next, the (V_{DS}, V_{GSE}) plane is optimally partitioned into a number of regions and a quadratic model of I_{DS} is numerically fitted in terms of V_{DS} and V_{GSE} in each region using the data points in that region. One example of the partitioned (V_{DS}, V_{GSE}) plane is shown in Fig. 3.3.

For a given regionwise partition, the following quadratic model of I_{DS} is fitted to the data in the kth region,

$$\begin{aligned}\frac{I_{DS}}{\beta} &= (\alpha_0^{(k)} + \alpha_1^{(k)} V_{GSE} + \alpha_2^{(k)} V_{DS} + \alpha_3^{(k)} V_{GSE}^2 + \\ &\quad \alpha_4^{(k)} V_{GSE} V_{DS} + \alpha_5^{(k)} V_{DS}^2), \quad k = 1, \ldots, n_r \end{aligned} \tag{3.18}$$

where β is the MOS device transconductance as in Eq. (3.1), n_r is the number of regions chosen for best fitting, and α's are fitting parameters in the kth region. Again, the carrier mobility (μ_0) and the zero-bias threshold voltage (V_{T0}) are the two physical parameters that are temperature dependent in the RWQ drain current equation Eq. (3.18). In [10], the temperature-dependence is incorporated into the RWQ model. It was observed in [10] that the threshold voltage is much less temperature sensitive than the carrier mobility. Therefore, the temperature-dependent V_{T0} model in [10] follows the simple MOS Level 1 model given in Eq. (3.2) and Eq. (3.3). However, a new temperature-dependent

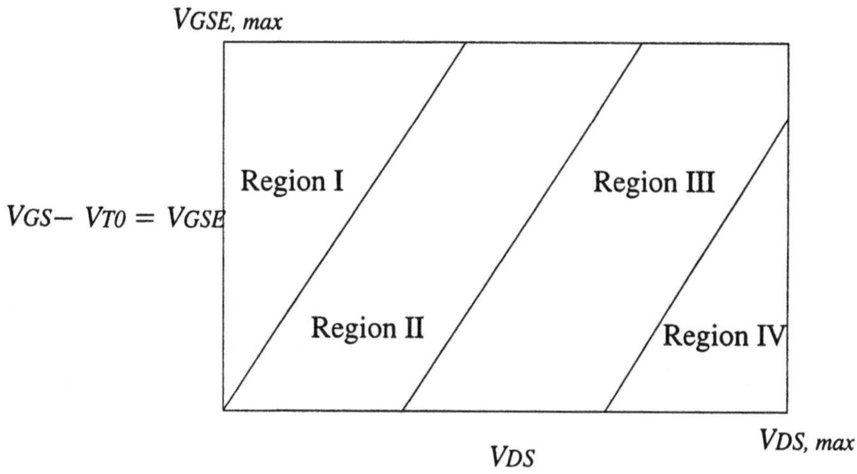

Figure 3.3. Regionwise partition of the (V_{DS}, V_{GSE}) plane.

μ_0 model that accurately take into account different scattering mechanisms has been developed in [10], which will be presented in the following.

3.4.1 TEMPERATURE-DEPENDENT MOBILITY MODELING

Because the on-chip temperature could be as high as 120 °C and the device feature size keeps shrinking in state-of-the-art VLSI technologies, the simple mobility model in Eq. (3.11) may not be sufficient to cover a wide range of temperatures. However, it is rather difficult to devise an analytical formula to accurately calculate the channel carrier mobility because of the complex quantum effects [11]. Although some empirical models such as SPICE BSIM use as many as eight fitting parameters to obtain a fairly good fit for the mobility of a fixed technology, the technology dependence and scaling properties are not well understood.

Owing to above reasons, a physically based, semiempirical mobility model has been developed [10] for the temperature range of 300 - 400 K, which is the normal temperature range for most circuits. Because this model is intended to be used in the fast-timing simulator, it must be accurate yet simple. In addition, this model should be scaled only with temperature and not with the transverse electric field E_{eff}, although the physical channel mobility indeed depends on both temperature and transverse electric field. This is because the transverse field dependence is already taken into account by V_{GSE} in Eq. (3.18) at the RWQ fitting stage, i.e., $E_{eff} \propto V_{GSE}$ [12].

Based on Eq. (3.6), Eq. (3.7), Eq. (3.9) and Eq. (3.10), the following temperature-dependent mobility model was proposed [10]:

$$\begin{aligned} U(T) &= \mu_0^{-1}(T) \\ &= a_1 T^{-1} + a_2 T^{a_3} + a_4 \\ &= A_1[(\frac{T}{300})^{-1} - 1] + A_2[(\frac{T}{300})^{A_3} - 1] + A_4. \end{aligned} \quad (3.19)$$

The symbol $U(T)$, defined as the inverse of the mobility $\mu_0(T)$, is used for convenience. The A_1, A_2, A_3 and A_4 terms are fitting parameters, which are determined by using the nonlinear least-square fitting technique to match the extracted $\mu_0(T)$. Method used to extract $\mu_0(T)$ will be given in the next section. The E_{eff} dependence in Eq. (3.7) and (3.9) is merged into A_2 and A_4 in Eq. (3.19).

3.4.2 $\mu_0(T)$ EXTRACTION FOR RWQ MODELING

Let $I_{DS}^{(k)}(\mathbf{x})$ and $\mu_0(T)f^{(k)}(\mathbf{x})$ be the experimental and RWQ-fitted drain currents in the k-th region, respectively. Here $\mathbf{x}$ denotes the data point vector (V_{DS}, V_{GSE}). The optimized mobility that produces the best fit can be extracted by minimizing the following objective function

$$\sum_{k=1}^{n_r} \sum_{i=1}^{N_k} (I_{DS}^{(k)}(\mathbf{x}_i) - \mu_0(T) f^{(k)}(\mathbf{x}_i))^2 \quad (3.20)$$

where n_r is the number of regions and N_k is the number of data points in region k. The minimization process provides us with a best overall RWQ fit to $I_{DS}(\mathbf{x})$ from which $\mu_0(T)$ is extracted to be [10]

$$\mu_0(T) = \frac{\sum_{k=1}^{n_r} \sum_{i=1}^{N_k} I_{DS}^{(k)}(\mathbf{x}_i) f^{(k)}(\mathbf{x}_i)}{\sum_{k=1}^{n_r} \sum_{i=1}^{N_k} (f^{(k)}(\mathbf{x}_i))^2}. \quad (3.21)$$

Once $\mu_0(T)$ values are extracted at several temperatures, Eq. (3.19) is used to find the optimized A_1, A_2, A_3 and A_4.

3.4.3 MOBILITY AND RWQ FITTING EXAMPLES

To evaluate the mobility modeling procedure described by Eq. (3.19) - Eq. (3.21), the values of $\mu_0(T)$ for an NMOSFET with specific dimensions, in this case, L = 0.8 μm and W = 1.6 μm, are extracted. Nonlinear least-square fitting of the A_1, A_2, A_3, and A_4 parameters is accomplished by using the Levenberg-Marquart algorithm [13]. The results are shown in Fig. 3.4 and the extracted parameters are given in the inset.

The RWQ modeling result at 27 °C is compared with the measured data in Fig. 3.5(a). The mobility model in Eq. (3.19) was used to predict $\mu_0(T =$

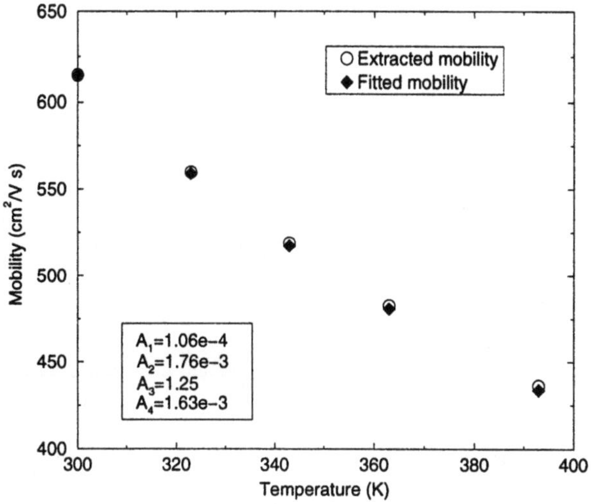

Figure 3.4. Fitted $\mu_0(T)$ vs. extracted $\mu_0(T)$.

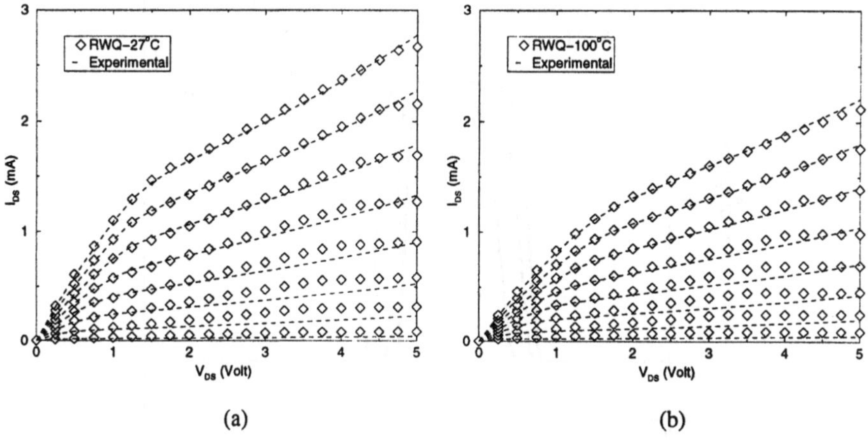

Figure 3.5. NMOSFET: (a) RWQ fitting result at 27 °C, and (b) RWQ fitting result at 100 °C with mobility optimization.

100 °C) as 464.2 $\mathrm{cm}^2/(\mathrm{V}\cdot\mathrm{s})$, which is very close to 466.3 $\mathrm{cm}^2/(\mathrm{V}\cdot\mathrm{s})$ obtained by extraction. The I_{DS}-V_{DS} curves at $T = 100$ °C were constructed by using the RWQ fitting parameters α_0 - α_5 obtained at room temperature and the value of $\mu_0(T = 100\ °\mathrm{C})$. The results are compared with the measured data in Fig. 3.5(b). Similarly, the RWQ fitting results for a PMOSFET with the same dimensions are shown in Fig. 3.6.

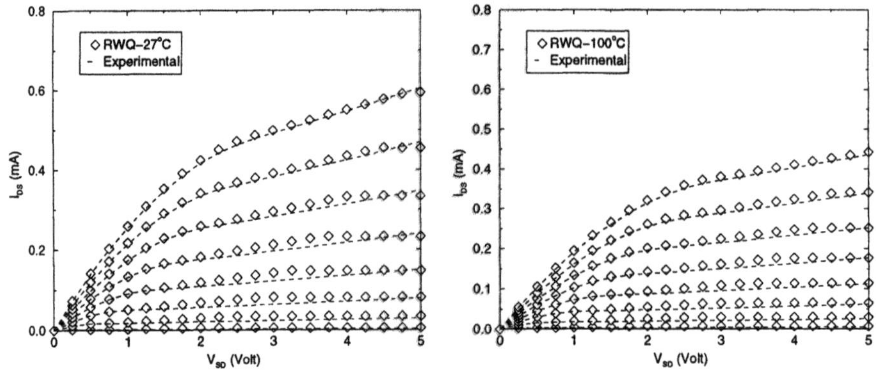

(a) (b)

Figure 3.6. PMOSFET: (a) RWQ fitting result at 27 °C, and (b) RWQ fitting result at 100 °C with mobility optimization.

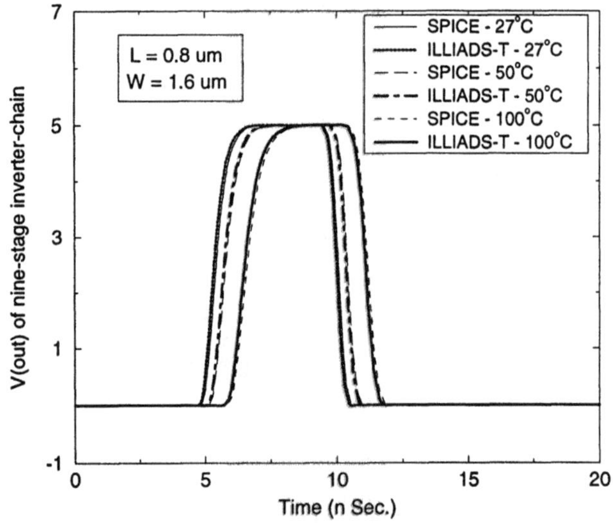

Figure 3.7. Waveform comparison between SPICE and ILLIADS-T for a 9-stage inverter chain.

Finally, to demonstrate the simulation accuracy of the fast-timing electrothermal simulator (ILLIADS-T)[10] in which the temperature-dependent RWQ model is implemented, a nine-stage inverter chain was simulated using both ILLIADS-T and SPICE with BSIM MOSFET models. The output waveforms at different temperatures are compared in Fig. 3.7.

3.5. SUMMARY

This chapter provides the temperature-dependent MOS device modeling techniques for circuit simulation.

- The temperature-dependent models of the threshold voltage and mobility of the MOS device are derived.

- To derive the temperature-dependent mobility model, the temperature dependence of three scattering mechanisms that determine the carrier mobility in semiconductors are introduced. The three scattering mechanisms are:

  - Coulomb (impurity) scattering
  - Lattice (phonon) scattering
  - Surface-roughness scattering

  Because the probability of a collision taking place in unit time is the sum of the probabilities of collision due to the scattering mechanisms, the carrier mobility can be expressed as the summation of the scattering-mechanism-limited mobilities.

- The temperature-dependent BSIM model for SPICE simulation is presented.

  - The general formula for describing the electrical parameters in BSIM is provided.
  - Since the number of electrical parameters in BSIM is large, the sensitivity analysis approach is used so that only sensitive parameters are analyzed when temperature changes.
  - In order to determine how sensitive the sensitive parameters are to the temperature variation, the temperature coefficients are determined by either measured experimental data or by analytically derived formula.

- The temperature-dependent regionwise-quadratic (RWQ) model for fast timing simulation is presented.

  - The working principles of the RWQ model are described.
  - The temperature-dependent mobility formula used in the RWQ model is developed. It is based on three scattering mechanisms and a number of fitting parameters. The fitting parameters are determined by the experimental data and the nonlinear least-square fit.
  - The simulation results are provided and compared to the measured data.

References

[1] H. Shichman and D. A. Hodges, "Modeling and simulation of insulated gate field effect transistor switching circuits," *IEEE Journal of Solid-State Circuits*, vol. 3, pp. 245–259, Sept. 1968.

[2] C. T. Sah, T. H. Ning, and L. L. Tschopp, "The scattering of electrons by surface oxide charges and by the lattice vibrations at the Si-SiO_2 interface," *Surface Science*, vol. 32, pp. 561–575, 1972.

[3] H. Ezawa, S. Kawaji, and K. Nakamura, "Surfons and the electron mobility in silicon inversion layers," *Japanese Journal of Applied Physics*, vol. 13, pp. 126–155, Sept. 1974.

[4] A. Hartstein, T. H. Ning, and A. B. Fowler, "Electron scattering in silicon inversion layers by oxide and surface roughness," *Surface Science*, vol. 58, pp. 181–190, 1976.

[5] C. P. Wan and B. J. Sheu, "Temperature dependence modeling for MOS VLSI circuit simulation," *IEEE Transactions on Computer-Aided Design of Integrated Circuits and Systems*, vol. 8, pp. 1065–1073, Oct. 1989.

[6] M. C. Hsu and B. J. Sheu, "Inverse-geometry dependence of MOS transistor electrical parameters," *IEEE Transactions on Computer-Aided Design of Integrated Circuits and Systems*, vol. 6, pp. 582–585, July 1987.

[7] P. Tomovic, *Sensitivity Analysis of Dynamic Systems.* New York, NY: McGraw-Hill, 1963.

[8] S. M. Sze, *Physics of Semiconductor Devices.* New York, NY: 2nd ed. John Wiley & Sons, 1981.

[9] A. Dharchoudhury, S. M. Kang, K. H. Kim, and S. H. Lee, "Fast and accurate timing simulation with regionwise quadratic models of MOS I-V characteristics," in *Proceedings of the ACM/IEEE International Conference on Computer-Aided Design*, pp. 208–211, Nov. 1994.

[10] Y. K. Cheng, P. Raha., C. C. Teng, E. Rosenbaum, and S. M. Kang, "ILLIADS-T: An electrothermal timing simulator for temperature-sensitive reliability diagnosis of CMOS VLSI chips," *IEEE Transactions on Computer-Aided Design of Integrated Circuits and Systems*, vol. 17, pp. 668–681, Aug. 1998.

[11] M. S. Lin, "A better understanding of the channel mobility of Si MOS-FET's based on the physics of quantized subbands," *IEEE Transactions on Electron Devices*, vol. 35, pp. 2406–2411, 1988.

[12] M. S. Liang, J. Y. Choi, P. K. Ko, and C. Hu, "Inversion-layer capacitance and mobility of very thin gate-oxide MOSFET's," *IEEE Transactions on Electron Devices*, vol. 33, pp. 409–412, 1986.

[13] D. W. Marquart *Journal of the Society for Industrial and Applied Math.*, vol. 11, pp. 431–441, 1963.

Chapter 4

THERMAL SIMULATION FOR VLSI SYSTEMS

4.1. INTRODUCTION

A thermal simulator finds the temperature distribution of a system containing heat sources during the electrothermal simulation, as depicted in Fig. 1.2. A heat source in a VLSI system may be a chip in a multi-chip module (MCM), or may simply be a heat dissipating device (e.g., transistor, resistor). Because the number of heat sources in a VLSI system can be large, the thermal simulation method for such a system must take into account the tradeoff between efficiency and accuracy.

The heat diffusion equation is the governing equation for describing the heat conduction and for calculating the temperature. The general equation is given in the following form [1]

$$\rho c_p \frac{\partial T(x,y,z,t)}{\partial t} = \nabla \cdot [k(x,y,z,T)\nabla T(x,y,z,t)] + g(x,y,z,t) \quad (4.1)$$

subject to the general thermal boundary condition

$$k(x,y,z,T)\frac{\partial T(x,y,z,t)}{\partial n_i} + h_i T(x,y,z,t) = f_i(x,y,z,t) \quad (4.2)$$

and the initial temperature condition

$$T(x,y,z,t=0) = F(x,y,z). \quad (4.3)$$

In Eq. (4.1) and Eq. (4.2), T is the temperature (°C), g is the power density of the heat source(s) (W/m^3), k is the thermal conductivity (W/(m°C)), ρ is the density of material (Kg/m^3), c_p is the specific heat (J/(Kg°C)), h_i is the heat transfer coefficient (W/(m^2 °C)), $f_i(x,y,z)$ is an arbitrary function, and n_i is the outward direction normal to the surface i. For the steady-state case, the $(\partial T/\partial t)$ term in Eq. (4.1) is zero.

If the thermal conductivity k is uniform (i.e., independent of position and temperature), Eq. (4.1) can be reduced to

$$\frac{1}{\alpha}\frac{\partial T(x,y,z,t)}{\partial t} = \nabla^2 T(x,y,z,t)] + \frac{g(x,y,z,t)}{k} \tag{4.4}$$

and Eq. (4.2) can be reduced to

$$k\frac{\partial T(x,y,z,t)}{\partial n_i} + h_i T(x,y,z,t) = f_i(x,y,z,t) \tag{4.5}$$

where α is called the thermal diffusivity of the medium defined as

$$\alpha = \frac{k}{\rho c_p}. \tag{4.6}$$

The following three types of thermal boundary conditions derived from Eq. (4.5) can be applied to the object boundaries, depending on the packaging materials and the surrounding environment:

$$\text{Isothermal (Dirichlet):} \quad T = f_i(x,y,z,t) \tag{4.7}$$

$$\text{Insulated (Neumann):} \quad \frac{\partial T}{\partial n_i} = 0 \tag{4.8}$$

$$\text{Convective (Robin):} \quad k\frac{\partial T}{\partial n_i} = h_i(T - T_a) \tag{4.9}$$

where T_a is the ambient temperature.

To solve the above boundary value problem of heat conduction, many existing analytical or numerical approaches are available. However, due to the large problem size in VLSI systems, some modification, simplification, or enhancement on those approaches are necessary. In Fig. 4.1, a mixed use of different thermal simulation methods that is particularly useful for a VLSI chip is illustrated [2]. This framework is comprised of three simulation engines: a fast thermal simulator, a numerical thermal simulator, and an analytical thermal simulator. The simulated on-chip substrate temperature profile is further utilized for interconnect temperature estimation as shown in Fig. 4.1. The interconnect temperature plays an important role in the RC delay calculation and electromigration reliability diagnosis, which will be discussed in Chapter 6 and Chapter 8.

The fast thermal simulator in the framework is designed to quickly identify the on-chip hot spots. It is very fast and provides a qualitative temperature description of the heat sources. The numerical thermal simulator is designed for the full-chip temperature profiling, while the analytical thermal simulator is used for accurately pinpointing the temperatures of the hot spots. Both numerical and analytical thermal simulation approaches take into account the

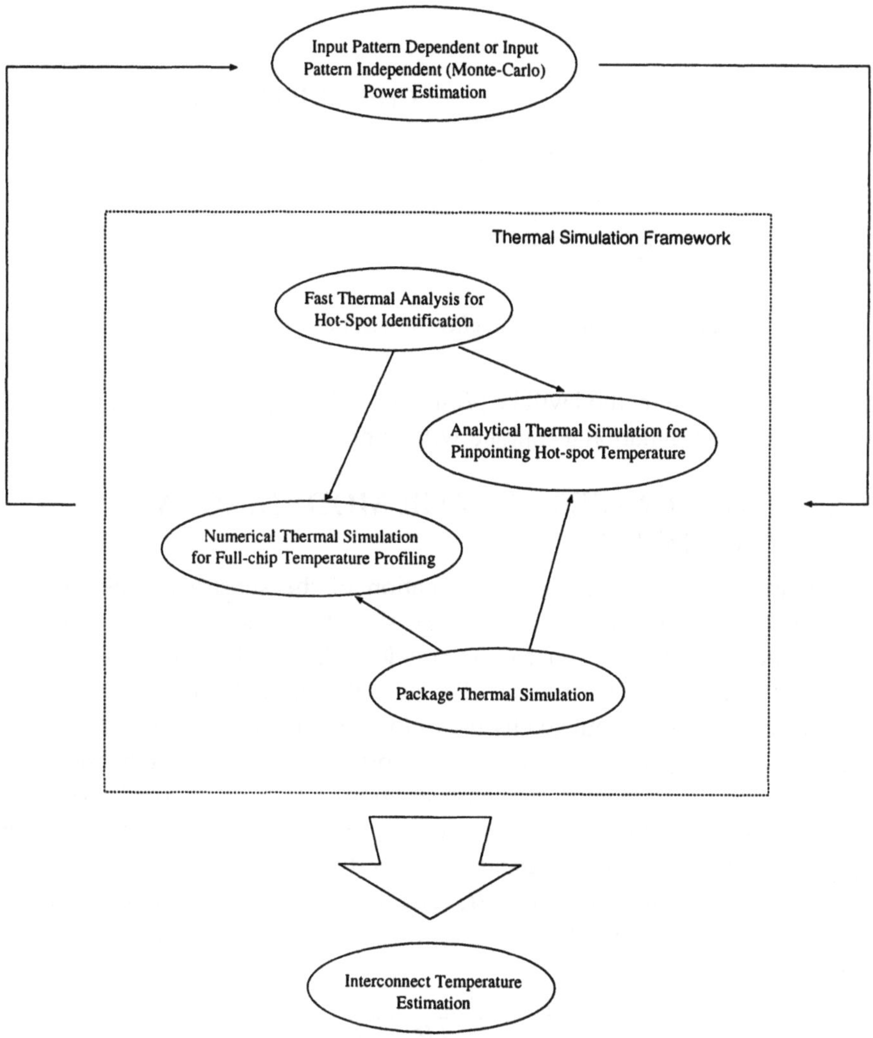

Figure 4.1. A thermal simulation framework [2].

packaging effect. Note that although the framework in Fig. 4.1 was originally developed to solve the electrothermal problem for a chip, it can be well applied to other VLSI systems.

In the following, a general description of the simulation strategy for a simple substrate/package structure of a chip will be given. Concepts such as the *thermal resistance* and the *effective heat transfer* will also be introduced. The fast thermal analysis, the numerical thermal simulation method, and the analytical

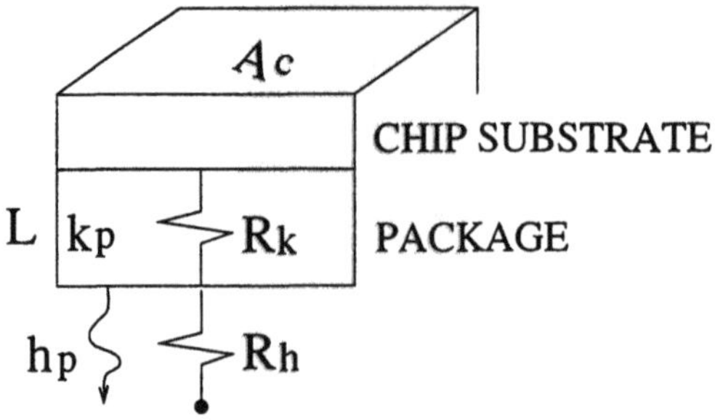

Figure 4.2. Illustration of effective heat transfer macromodeling.

thermal simulation method will be presented and compared. Finally, the details of the package simulation technique will be presented.

4.2. SUBSTRATE/PACKAGE MODELING: AN OVERVIEW

A chip can be viewed as a combination of the substrate (bulk) and the package. A package consists of the heat sink, the pins, and many other heat removal devices. An exact approach to model a chip for thermal simulation is to directly model the whole structure. Because of its complexity, however, a chip is often modeled and simulated by the following mixed 3-D/1-D strategies: (1) 3-D simulation is performed for the chip substrate to achieve a higher degree of accuracy, and (2) packaging elements are modeled as 1-D thermal resistances to reduce the computation cost. The strategy (2) is henceforth referred to as the *effective heat transfer macromodeling.* Specifically, as shown in Fig. 4.2, the thermal resistance of the package (R_k) and the thermal resistance from package to ambient (R_h) are serially combined to calculate the effective heat transfer coefficient h^e given by

$$h^e = \frac{1}{A_c} \frac{1}{(R_h + R_k)}. \tag{4.10}$$

In Eq. (4.10), $R_k = \frac{L}{k_p A_c}$, $R_h = \frac{1}{h_p A_c}$, L and k_p are the thickness and the thermal conductivity of the package respectively, h_p is the heat transfer coefficient from package to ambient, and A_c is the chip area normal to the direction of heat flow. From Eq. (4.10) it can be seen that the package effects are merged into the h_i term in Eq. (4.2) and an effective h^e is formed. The advantage of the effective heat transfer macromodeling is threefold. First, it improves the overall simulation efficiency. Second, it decouples the thermal simulation problem between the substrate and the package, so that Eq. (4.2) remains valid with the exception that h_i is replaced by h^e. Also the substrate and the package can be characterized and simulated independently. Third, it allows the complicated chip structures to be easily handled. For instance, if there are pins existing in Fig. 4.2, they can be taken into account by replacing k_p in R_k with

$$k_{eff} = Xk_{pin} + (1 - X)k_p \tag{4.11}$$

where k_{pin} is the thermal conductivity of the pins and

$$X = \frac{\text{(Area of pins)}}{\text{(Total package area)}}. \tag{4.12}$$

4.3. FORMULATION OF THERMAL ANALYSIS

The different thermal simulation methods in the framework shown in Fig. 4.1 can be utilized to generate the temperature profile, identify the hot spots, and pinpoint the hot-spot temperatures for the VLSI systems. In the following, the formulation of each thermal simulation technique in the framework will be presented. The advantages and disadvantages of each technique will be also discussed.

4.3.1 FAST THERMAL ANALYSIS

For a VLSI system containing a large number of heat sources, the exact numerical and analytical methods are computationally inefficient. In the early design phase when no specific package information is given or the thermal boundary condition is not fully characterized, a fast thermal analysis method that emphasizes the hot-spot identification is highly desirable. It can be used for the iterative temperature-sensitive module placement in order to achieve a more uniform temperature distribution for better reliability and reduced delay. In this section, a fast thermal analysis method, called FTA [3], is introduced. This method has been shown to quickly identify the on-chip hot spots. Several assumptions were made in [3] specifically for a VLSI chip, therefore the presented formulation may need to be modified to suit certain needs for other environment. However, the basic concept remains the same.

The FTA approach utilizes the fact that the dimensions of the gate-level or subcircuit-level heat sources in a VLSI chip are small compared with the chip size. Therefore, all heat sources can be viewed as located in a virtually *infinite* body. Consider a point source in a chip as shown in Fig. 4.3(a). Since ICs have a passivation layer, the top of the chip is insulated. We therefore have a boundary value problem with infinite dimension in the x-y plane while with semi-infinite dimension in the z direction. Moreover, the boundary condition at $z = 0$ is $\frac{\partial T(\mathbf{r},t)}{\partial z} = 0$. To find the temperature subject to this specific geometry and boundary conditions, the *method of images* is used [3]: an identical heat source that is symmetric with respect to $z = 0$ is added and the insulating boundary is removed. Now, the problem in Fig. 4.3(a) is transformed to that in Fig. 4.3(b).

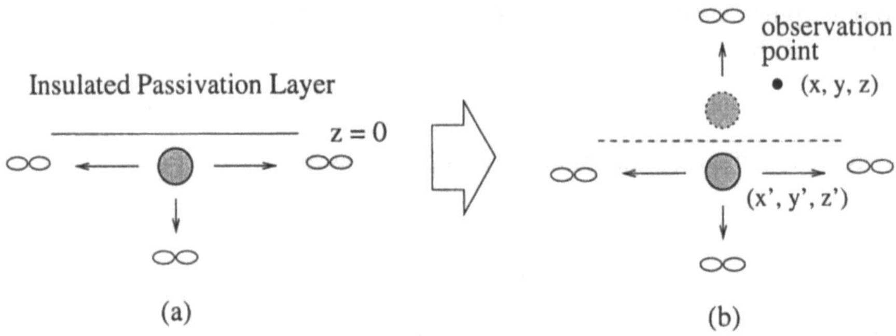

Figure 4.3. Method of images.

The Green's function solution $G(\mathbf{r}, t|\mathbf{r}', \tau)$ of the heat diffusion equation Eq. (4.4) for the point source in Fig. 4.3 can be derived as $G(\mathbf{r}, t|\mathbf{r}', \tau) = G_x \cdot G_y \cdot G_z$, where

$$\begin{aligned} G_x &= \frac{1}{[4\pi\alpha(t-\tau)]^{1/2}} \exp[-\frac{(x-x')^2}{4\alpha(t-\tau)}] \\ G_y &= \frac{1}{[4\pi\alpha(t-\tau)]^{1/2}} \exp[-\frac{(y-y')^2}{4\alpha(t-\tau)}] \\ G_z &= \frac{1}{[4\pi\alpha(t-\tau)]^{1/2}} [\exp(-\frac{(z-z')^2}{4\alpha(t-\tau)}) + \exp(-\frac{(z+z')^2}{4\alpha(t-\tau)})] \end{aligned} \quad (4.13)$$

where α is the thermal diffusivity. The resulting temperature rise above the ambient at observation point $\mathbf{r}$ due to the parallelepiped heat source with dimensions $a \times b \times c$ can be formulated as

$$\Delta T(\mathbf{r}, t) = \frac{\alpha P_0}{k(abc)} \int_{\tau=0}^{t} \int_{-c}^{c} \int_{-b/2}^{b/2} \int_{-a/2}^{a/2} G(\mathbf{r}, t|\mathbf{r}', \tau) dv' d\tau \quad (4.14)$$

where the coordinate origin has been set to be at the center of the source and P_0 is the source power.

By utilizing the error function, Eq. (4.14) can be simplified to:

$$\Delta T(x, y, 0, t) = \frac{\alpha P_0}{k(abc)} \int_0^t G(x, a, \tau) \cdot G(y, b, \tau) \cdot G(0, c, \tau) d\tau \quad (4.15)$$

where

$$\begin{aligned} G(x, a, \tau) &= \frac{1}{2}[\text{erf}(\frac{a/2+x}{2\sqrt{\alpha(t-\tau)}}) + \text{erf}(\frac{a/2-x}{2\sqrt{\alpha(t-\tau)}})] \\ G(y, b, \tau) &= \frac{1}{2}[\text{erf}(\frac{b/2+y}{2\sqrt{\alpha(t-\tau)}}) + \text{erf}(\frac{b/2-y}{2\sqrt{\alpha(t-\tau)}})] \end{aligned}$$

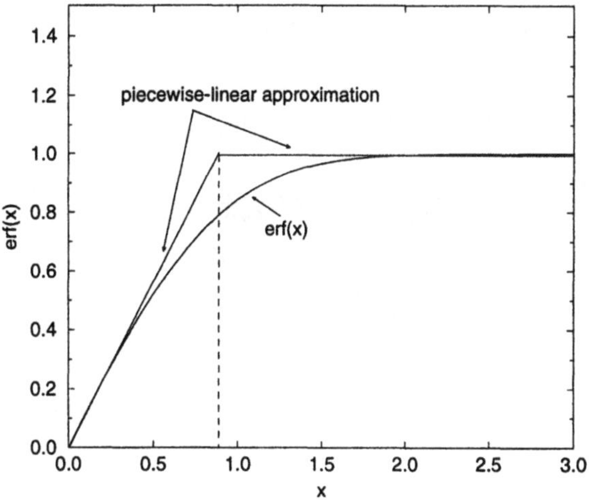

Figure 4.4. Error function approximation.

$$G(0, c, \tau) = \text{erf}(\frac{c}{2\sqrt{\alpha(t-\tau)}}). \tag{4.16}$$

In Eq. (4.15), the observation point is set to be on the chip surface by replacing **r** with $(x, y, 0)$. Defining $\mathcal{A}_1 = 2(a/2+x)$, $\mathcal{A}_2 = 2(a/2-x)$, $\mathcal{B}_1 = 2(b/2+y)$, $\mathcal{B}_2 = 2(b/2-y)$, and $\mathcal{C} = 2c$, along with the change of variables, Eq. (4.15) can be rewritten as

$$\begin{aligned}\Delta T(x, y, 0, t) = & \frac{\alpha P_0}{4k(abc)} \int_0^t [\text{erf}(\frac{\mathcal{A}_1}{4\sqrt{\alpha\tau}}) + \text{erf}(\frac{\mathcal{A}_2}{4\sqrt{\alpha\tau}})] \cdot \\ & [\text{erf}(\frac{\mathcal{B}_1}{4\sqrt{\alpha\tau}}) + \text{erf}(\frac{\mathcal{B}_2}{4\sqrt{\alpha\tau}})] \cdot \text{erf}(\frac{\mathcal{C}}{4\sqrt{\alpha\tau}}) d\tau. \end{aligned} \tag{4.17}$$

In order to perform the integration in Eq. (4.17) analytically, the error function is piecewise linearized according to (referring to Fig. 4.4) [4]

$$\text{erf}(x) \approx \begin{cases} 2x/\sqrt{\pi} & \text{if } x \leq \sqrt{\pi}/2 \\ 1 & \text{if } x > \sqrt{\pi}/2. \end{cases} \tag{4.18}$$

Table 4.1 demonstrates the approximation results by using Eq. (4.18) and defining $t_{a1} = \mathcal{A}_1^2/(4\pi\alpha)$, $t_{a2} = \mathcal{A}_2^2/(4\pi\alpha)$, $t_{b1} = \mathcal{B}_1^2/(4\pi\alpha)$, $t_{b2} = \mathcal{B}_2^2/(4\pi\alpha)$, and $t_c = \mathcal{C}^2/(4\pi\alpha)$. In Table 4.1, $m = 0$ if $\mathcal{A}_2 > 0$, otherwise $m = 1$. Similarly, $n = 0$ if $\mathcal{B}_2 > 0$, otherwise $n = 1$. To obtain the analytical solution of Eq. (4.17), the number of possible permutations (i.e., 120) of t_{a1}, t_{a2}, t_{b1}, t_{b2} and t_c needs to be reduced to a manageable level. To do this, the following constraints are asserted according to [3]:

1. $x \geq 0, y \geq 0$

Table 4.1. Error function approximations.

$$\mathrm{erf}\left(\frac{A_1}{4\sqrt{\alpha\tau}}\right) \approx \begin{cases} \left(\frac{t_{a1}}{\tau}\right)^{1/2} & \text{if } \tau \geq t_{a1} \\ 1 & \text{if } \tau < t_{a1} \end{cases} \qquad \mathrm{erf}\left(\frac{A_2}{4\sqrt{\alpha\tau}}\right) \approx \begin{cases} (-1)^m\left(\frac{t_{a2}}{\tau}\right)^{1/2} & \text{if } \tau \geq t_{a2} \\ (-1)^m & \text{if } \tau < t_{a2} \end{cases}$$

$$\mathrm{erf}\left(\frac{B_1}{4\sqrt{\alpha\tau}}\right) \approx \begin{cases} \left(\frac{t_{b1}}{\tau}\right)^{1/2} & \text{if } \tau \geq t_{b1} \\ 1 & \text{if } \tau < t_{b1} \end{cases} \qquad \mathrm{erf}\left(\frac{B_2}{4\sqrt{\alpha\tau}}\right) \approx \begin{cases} (-1)^n\left(\frac{t_{b2}}{\tau}\right)^{1/2} & \text{if } \tau \geq t_{b2} \\ (-1)^n & \text{if } \tau < t_{b2} \end{cases}$$

$$\mathrm{erf}\left(\frac{C}{4\sqrt{\alpha\tau}}\right) \approx \begin{cases} \left(\frac{t_c}{\tau}\right)^{1/2} & \text{if } \tau \geq t_c \\ 1 & \text{if } \tau < t_c \end{cases}$$

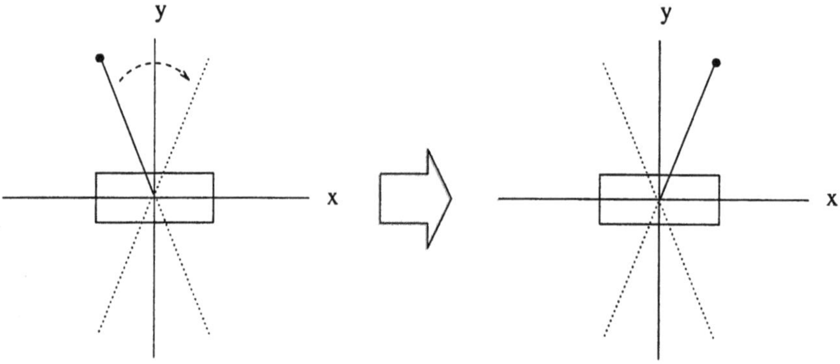

Figure 4.5. Transformation 1: Constrain the observation point to the first quadrant.

2. $(\frac{a}{2}+x) \geq |\frac{a}{2}-x| \Longrightarrow A_1^2 \geq A_2^2 \Longrightarrow t_{a1} \geq t_{a2}$

3. $(\frac{b}{2}+y) \geq |\frac{b}{2}-y| \Longrightarrow B_1^2 \geq B_2^2 \Longrightarrow t_{b1} \geq t_{b2}$

4. $2(\frac{a}{2}+x) \geq 2c \Longrightarrow A_1^2 \geq C^2 \Longrightarrow t_{a1} \geq t_c$

5. $2(\frac{b}{2}+y) \geq 2c \Longrightarrow B_1^2 \geq C^2 \Longrightarrow t_{b1} \geq t_c$

6. $t_{a1} \geq t_{b1}$

Constraints (2) and (3) are straightforward algebraically. Constraints (4) and (5) are valid because the thickness of the heat sources (i.e., c) is much smaller than the physical dimensions (i.e., a and b) of a logic gate. Constraint (1) is equivalent to transforming all the observation points to the first quadrant by using the symmetric property as graphically shown in Fig. 4.5. To satisfy Constraint (6), the coordinate transformation shown in Fig. 4.6 is used. With the above specified constraints, Eq. (4.17) now becomes tractable. In other words, the order of t_{a1}, t_{a2}, t_{b1}, t_{b2}, and t_c must belong to one of the eight cases shown in Fig. 4.2. The analytical solutions are derived for all cases. Proper solutions will be used during the simulation, depending on the geometry and

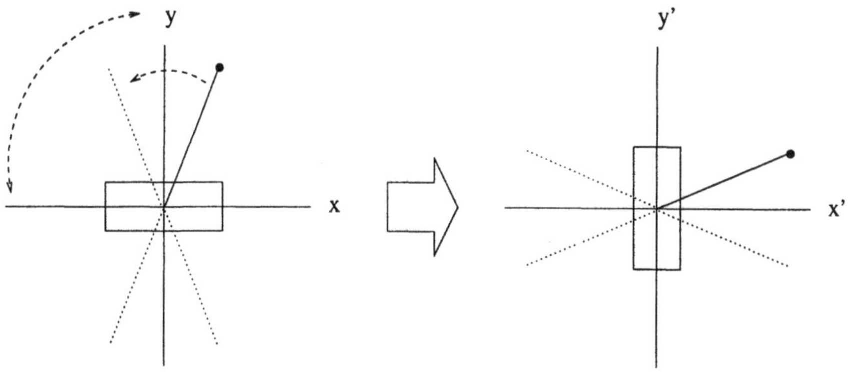

Figure 4.6. Transformation 2: Constrain t_{a1} to be larger than t_{b1}.

Table 4.2. Eight cases under six constraints.

Case 1. $t_{a1} \geq t_{b1} \geq t_{b2} \geq t_{a2} \geq t_c$	Case 2. $t_{a1} \geq t_{b1} \geq t_{a2} \geq t_{b2} \geq t_c$
Case 3. $t_{a1} \geq t_{a2} \geq t_{b1} \geq t_{b2} \geq t_c$	Case 4. $t_{a1} \geq t_{a2} \geq t_{b1} \geq t_c \geq t_{b2}$
Case 5. $t_{a1} \geq t_{b1} \geq t_{a2} \geq t_c \geq t_{b2}$	Case 6. $t_{a1} \geq t_{b1} \geq t_{b2} \geq t_c \geq t_{a2}$
Case 7. $t_{a1} \geq t_{b1} \geq t_c \geq t_{b2} \geq t_{a2}$	Case 8. $t_{a1} \geq t_{b1} \geq t_c \geq t_{a2} \geq t_{b2}$

the size of the heat sources, as well as the relative locations between the heat source and the observation point.

For a VLSI chip with n heat sources, the temperature rise at the center of source i is obtained by considering the heat diffusion from i itself, plus that from other $n-1$ sources using superposition:

$$\Delta T_i^{\Sigma} = \Delta T_i(\mathbf{0}, t) + \sum_{k=1}^{n-1} \Delta T_k(\mathbf{r}_k, t), \tag{4.19}$$

where ΔT_i^{Σ} is the temperature rise at the center of source i, $\Delta T_i(\mathbf{0}, t)$ is the temperature rise due to i itself, and $\Delta T_k(\mathbf{r}_k, t)$ is the temperature rise due to source k. Here, $\Delta T_i(\mathbf{0}, t)$ and $\Delta T_k(\mathbf{r}_k, t)$ can both be found by combining Eq. (4.17) with one of the eight cases in Fig. 4.2.

Let us consider Fig. 4.7 as an example, where one source with power P_I is located at (0,0) and the other one with power P_{II} is at (3,2). The temperature rise at the center of source II due to source I (i.e., $\Delta T_k(\mathbf{r}_k, t)$ in Eq. (4.19)) can be found by recognizing that this condition satisfies Case 1 above, where $t_{a1} \geq t_{b1} \geq t_{b2} \geq t_{a2} \geq t_c$. Thus the integration in Eq. (4.17) can be explicitly performed. Specifically, at steady state ($t \rightarrow \infty$):

$$\Delta T_I(x, y, 0, \infty) \quad = \quad \frac{\alpha P_I}{4k(abc)}[(\sqrt{t_{b1}t_c} - \sqrt{t_{b2}t_c}) \cdot (4 + 2\log(\frac{\mathcal{A}_1}{\mathcal{B}_1}))$$

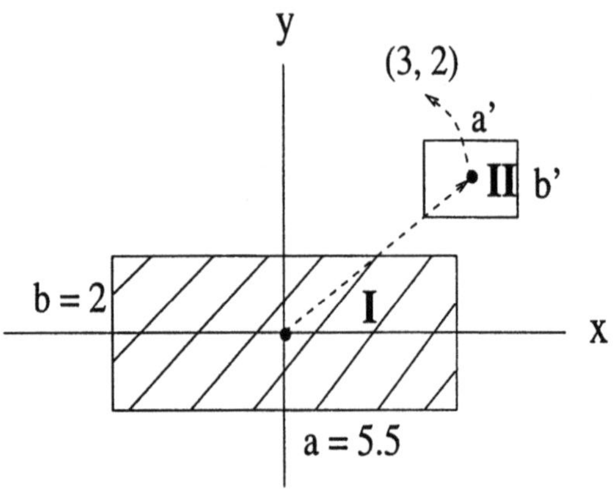

Figure 4.7. An FTA example.

$$-(\sqrt{t_{b2}t_c} + \sqrt{t_{a2}t_c}) \cdot 2\log(\frac{\mathcal{B}_1}{\mathcal{B}_2})] \qquad (4.20)$$

if $m = 1$ and $n = 1$ where $(x, y) = (3, 2)$. Similarly, the temperature rise at the center of source II due to self heating (i.e., $\Delta T_i(\mathbf{0}, t)$ in Eq. (4.19)) can be found since it is a special case ($t_{a1} = t_{a2} \geq t_{b1} = t_{b2} \geq t_c$) of Case 3:

$$\Delta T_{II}(\mathbf{0}, \infty) = \frac{P_{II}}{\pi k a'}[2 + \log(\frac{a'}{b'}) - \frac{c}{b'}]. \qquad (4.21)$$

Finally, the steady-state temperature rise at the center of source II can be obtained, according to Eq. (4.19), as

$$\Delta T_{II}^{\Sigma} = \Delta T_{II}(\mathbf{0}, \infty) + \Delta T_I(3, 2, 0, \infty). \qquad (4.22)$$

The mathematical formulation of the FTA method is based on the closed-form Green's function with the assumption of semi-infinite boundary condition. In practice, this assumption is not valid due to the existence of package and heat sink in a chip. Therefore, the temperature rise predicted by Eq. (4.19) represents the relative value instead of the absolute value. However, the FTA method provides a quick estimate of the temperature distribution. This is particularly useful when the number of heat sources is large or when a large number of repeated thermal simulations are needed. It is also useful when the chip package specification is not known in the early design phase. In order to take into account the effects of package and heat sink, detailed thermal simulation using a numerical or analytical approach is needed.

Among different thermal simulation methods, the FTA method is primarily used for fast hot-spot identification. In order to observe how accurately the

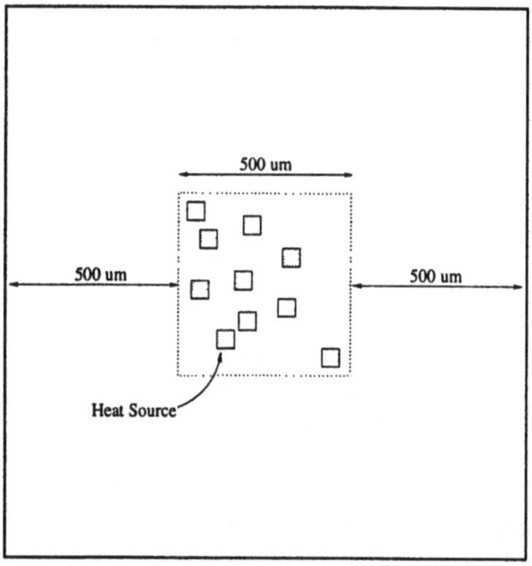

Figure 4.8. Chip structure and heat source locations.

Table 4.3. Violation rate by using the FTA method.

h [W/m^2 K]	5,000	8,000	10,000	15,000
# violation	4	5	5	4
Violation rate (%)	8	10	10	8
ΔT_{hot_spot} [°C]	1.1	0.83	0.6	0.8

FTA method can identify the hot spots, the following experiment is performed. Consider a chip containing 10 heat sources, all with dimensions of 50μm $\times$ 50μm. The sources are confined within the area (source area) with dimensions of 500μm $\times$ 500μm, and the distance between the boundary of the source area and the chip's bonding pad is 500μm as shown in Fig. 4.8. The heat sources are randomly placed and the power values from 10 mW to 100 mW are randomly assigned to the sources. The heat source with the highest temperature is found by using both the FTA method and the exact numerical thermal simulation method. The above process is repeated for 50 times (i.e., 50 tests). A violation happens in a test if the hot spot identified by the FTA method is different from the one identified by the exact method.

The number of violation and the violation rate among 50 tests are shown in the second and third rows of Table 4.3 for different h (heat transfer coefficient in Eq. (4.2)) values. Note that the exact method needs the h value in order to take the boundary conditions into account. In Table 4.3, ΔT_{hot_spot} is defined

Table 4.3 (continued). Violation rate by using the FTA method.

h [W/m^2 K]	20,000	25,000	30,000	35,000
# violation	4	6	3	3
Violation rate (%)	8	12	6	6
ΔT_{hot_spot} [°C]	0.35	0.75	0.21	0.16

as

$$\Delta T_{hot\_spot} = \frac{1}{\#\,\text{violation}} \sum_{i=1}^{\#\,\text{violation}} \left| T^i_{exact} - T^i_{FTA} \right| . \tag{4.23}$$

Here, T^i_{exact} is the actual temperature of the hot spot that is identified by the exact method, and T^i_{FTA} is the actual temperature of the hot spot that is identified by the FTA method. It shows that a violation occurs only when the two hot spots have very close temperature values, which is insignificant for reliability concerns. Therefore, the FTA method proposed in [3] is generally accurate in finding the hot spots in the VLSI chip. The efficiency of the FTA method will be presented later in the chapter, after the exact numerical and analytical simulation methods are introduced.

4.3.2 NUMERICAL APPROACH

The heat conduction problem can be solved numerically by using the finite-difference method (FDM) [5, 6, 7], the finite-element method (FEM) [8, 9, 10], or the boundary-element method (BEM) [11] among others. In this section, the FDM will be discussed and it will remain to be the main numerical method used for illustration throughout this book.

The basic concept of the finite-difference approach is to approximate the partial derivative of a given point by a derivative taken over a finite interval across that point. Let $f(x)$ be a function which is finite, continuous and single-valued. The derivative of $f(x)$ at point x_i can be approximated by the following difference equation:

$$\left.\frac{df(x)}{dx}\right|_{x_i} = \frac{f(x_i + h) - f(x_i - h)}{2h} \tag{4.24}$$

where $(x_i + h)$ and $(x_i - h)$ are the two neighboring points. Similarly, the finite-difference approximation of the second derivative of $f(x)$ can be given as

$$\left.\frac{d^2 f(x)}{dx^2}\right|_{x_i} = \frac{f(x_i + h) + f(x_i - h) - 2f(x_i)}{h^2}. \tag{4.25}$$

The above finite-difference expressions of the first and second derivatives can be applied to Eq. (4.1) or Eq. (4.4) with respect to the time domain and the space domain. The object under simulation is first discretized into many space grid points x_i, and the algebraic difference equations are obtained for each x_i. The resulting set of equations that represent all grid points can be solved for the successive time points provided that the temperature distribution at $t = 0$ is given.

The number of grids and the distance between two grid points determine the accuracy and simulation speed of the FDM. Furthermore, the distance between the grid points (x_i,x_{i+1}) can be different from the distance between the grid points (x_{i+1},x_{i+2}). To account for both accuracy and speed, an *adaptive meshing* technique is often employed in the FDM to optimally generate the variable grid system [6]. The grid lines are first uniformly deployed according to the user-specified initial number of grids. After an initial estimate of the temperature distribution is obtained, the grids are refined or redistributed by sensing the temperature gradient. Extra grids are added to the regions with larger gradient based on the following *weight function* (Eq. (4.26)) and *equidistribution* criterion (Eq. (4.27)) [12]:

$$w(r) = \sqrt{1 + \alpha^2 (\frac{\partial T}{\partial r})^2} \tag{4.26}$$

$$\int_{r_i}^{r_{i+1}} w(r)dr = \text{Constant} \tag{4.27}$$

where α is the user-specified tuning parameter and r denotes the x or y axis. It can be seen that for a larger gradient, a smaller grid spacing is needed. Temperature solutions of the grids in the previous grid system are compared to those in the refined grid system. If the percentage difference is less than a prescribed threshold, then the grid refinement process is terminated. The stopping criterion can also be the user-specified maximum number of grids.

To start deriving the finite-difference equations similar to Eq. (4.24) and Eq. (4.25) for the 3-D Cartesian coordinate system, the schematic representation of a solid containing several heat sources along with the variable grid system is given in Fig. 4.9. After the coordinates of each heat source are identified, the corresponding grid points into which the heat flows and the proportionate power values in the analogous thermal circuit are found as shown in Fig. 4.10. Symbol P_i in Fig. 4.10(a) denotes the heat flow from source i. The solid lines in Fig. 4.9(a) and Fig. 4.10(a) represent the chosen grid lines and dashed ones are in the middle of two adjacent grid lines. A_{eff} is the effective area of a grid point.

Every heat source that overlaps the effective area of a grid point serves as a power source feeding into that grid, and the corresponding power value is calculated based on the ratio of the source area within A_{eff} to the total area of the source. In Fig. 4.9(b), h_{x+}, h_{x-}, h_{y+}, h_{y-}, h_{z+} and h_{z-} are

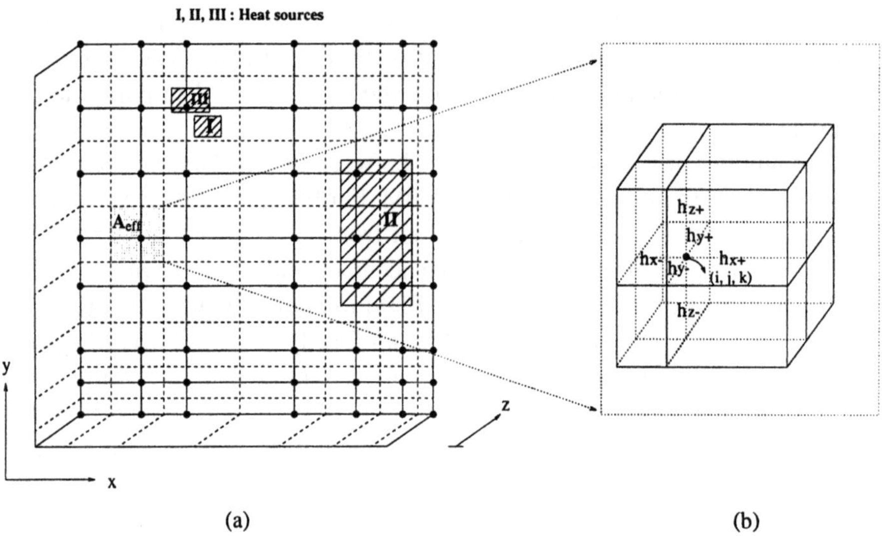

Figure 4.9. (a) Top view of the solid containing heat sources, and (b) 3-D view of grid point (i, j, k).

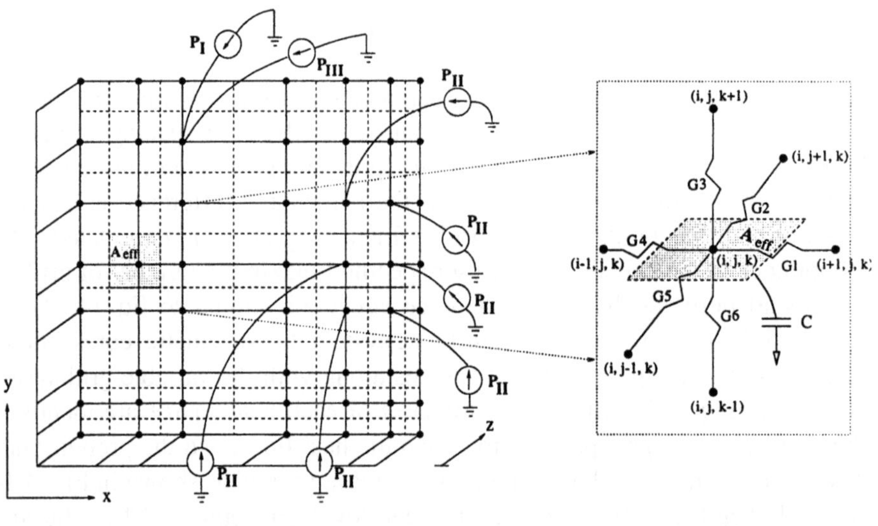

Figure 4.10. (a) Analogous thermal circuit to Fig. 4.9(a), and (b) thermal conductances from (i, j, k) to adjacent grids.

halves of the distances from grid (i, j, k) to grids $(i+1, j, k)$, $(i-1, j, k)$, $(i, j+1, k)$, $(i, j-1, k)$, $(i, j, k+1)$, and $(i, j, k-1)$, respectively. The thermal conductances G_1, G_2, G_3 and the thermal capacitance C in Fig. 4.10(b) can be

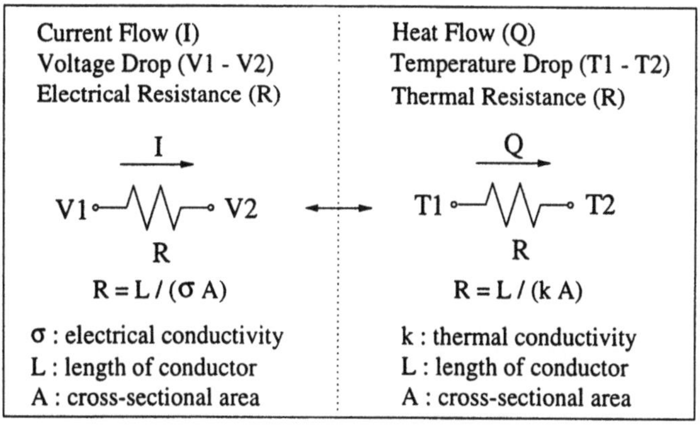

Figure 4.11. Analogy between thermal and electrical circuits.

found by applying the *first law of thermodynamics* (i.e., energy conservation in a thermodynamic system) on the grid point (i, j, k):

$$\begin{aligned}
&k(h_{y+}+h_{y-})(h_{z+}+h_{z-})\cdot\left[\frac{T^n_{i+1,j,k}-T^n_{i,j,k}}{2h_{x+}}+\frac{T^n_{i-1,j,k}-T^n_{i,j,k}}{2h_{x-}}\right]+\\
&k(h_{x+}+h_{x-})(h_{z+}+h_{z-})\cdot\left[\frac{T^n_{i,j+1,k}-T^n_{i,j,k}}{2h_{y+}}+\frac{T^n_{i,j-1,k}-T^n_{i,j,k}}{2h_{y-}}\right]+\\
&k(h_{x+}+h_{x-})(h_{y+}+h_{y-})\cdot\left[\frac{T^n_{i,j,k+1}-T^n_{i,j,k}}{2h_{z+}}+\frac{T^n_{i,j,k-1}-T^n_{i,j,k}}{2h_{z-}}\right]\\
&=(h_{x+}+h_{x-})(h_{y+}+h_{y-})(h_{z+}+h_{z-})\rho c_p\frac{T^{n+1}_{i,j,k}-T^n_{i,j,k}}{\Delta t} \qquad (4.28)
\end{aligned}$$

where Δt is the time increment and T^n is the temperature at time $n \cdot \Delta t$. From Eq. (4.28), it can be seen that the heat conduction in a thermal circuit is similar to the current conduction in an electrical circuit with the analogy shown in Fig. 4.11. Therefore, a finite-difference heat conduction problem can be mapped into an electrical RC network problem. In fact, Eq. (4.28) is analogous to the Kirchhoff's current law (KCL) used in the electrical circuit. The thermal conductances and capacitance connected to the grid (i, j, k) can thus be found as

$$\begin{aligned}
G_1 &= \frac{k\cdot(h_{y+}+h_{y-})(h_{z+}+h_{z-})}{2h_{x+}}\\
G_2 &= \frac{k\cdot(h_{x+}+h_{x-})(h_{z+}+h_{z-})}{2h_{y+}}
\end{aligned}$$

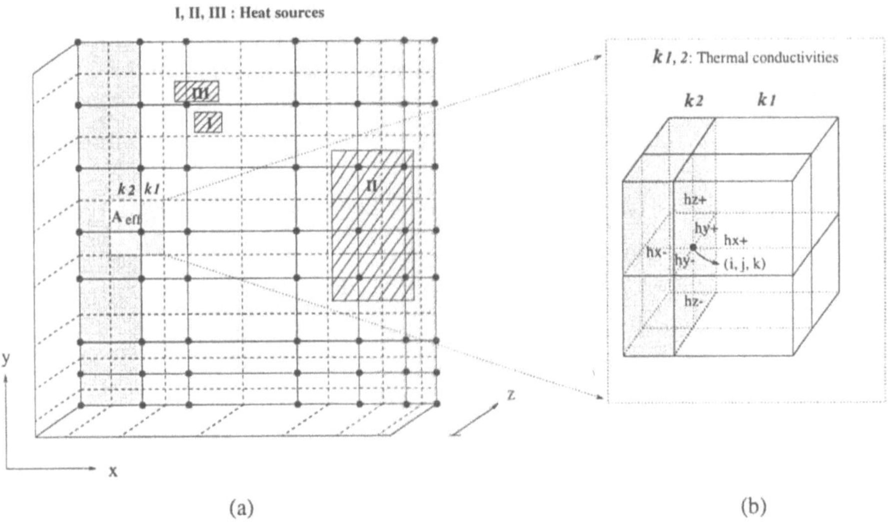

Figure 4.12. (a) Top view of a part of the chip comprised of composite materials, and (b) 3-D view of grid point (i, j, k).

$$
\begin{aligned}
G_3 &= \frac{k \cdot (h_{x+} + h_{x-})(h_{y+} + h_{y-})}{2h_{z+}} \\
C &= (h_{x+} + h_{x-})(h_{y+} + h_{y-})(h_{z+} + h_{z-})\rho c_p.
\end{aligned}
\tag{4.29}
$$

Similar expressions for G_4, G_5 and G_6 can be derived. For a composite material system, as shown in Figs. 4.12 and 4.13, G_1, G_2, and G_3 are found to be

$$
\begin{aligned}
G_1 &= \frac{k1}{2h_{x+}}(h_{y+} + h_{y-})(h_{z+} + h_{z-}) \\
G_2 &= \frac{k1 \cdot h_{x+} + k2 \cdot h_{x-}}{2h_{y+}}(h_{z+} + h_{z-}) \\
G_3 &= \frac{k1 \cdot h_{x+} + k2 \cdot h_{x-}}{2h_{z+}}(h_{y+} + h_{y-}).
\end{aligned}
\tag{4.30}
$$

The thermal circuit is represented by a large set of nodal equations like Eq. (4.28), and a system matrix is created. This matrix can be solved by either the sparse-matrix technique or the successive-over-relaxation (SOR) technique in order to obtain the temperature distribution of the object under simulation. The matrix is solved successively for each time increment Δt, based on the temperature values of the grids at the previous time point. In the electrothermal simulation flow for the VLSI chip, once the on-chip temperature profile is found, the gate temperature is calculated by averaging the temperature values of the grids that the gate covers.

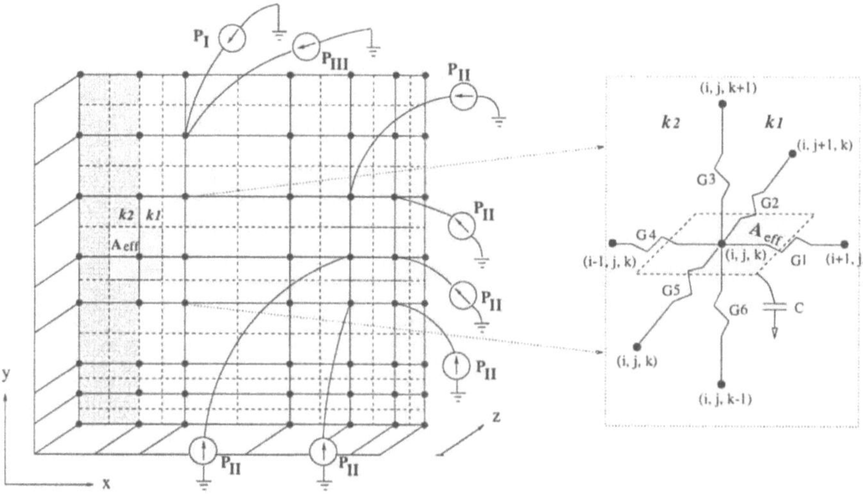

(a) (b)

Figure 4.13. (a) Analogous thermal circuit to Fig. 4.12(a), and (b) thermal conductances from (i, j, k) to adjacent grids.

The previous discussion describes the finite-difference temperature calculation for the interior grids. However, special care must be taken for the boundary grids. In the following, the 3-D steady-state case is illustrated without loss of generality. Figure 4.14 shows the steady-state thermal circuit used to model the top of a solid with the convective boundary condition, where T_a is the ambient temperature. To find the equivalent thermal conductances for this system, the first law of thermodynamics is again applied on the grid point (i, j, k):

$$
\begin{aligned}
&k(h_{y+} + h_{y-})h_{z-}\frac{T_{i+1,j,k} - T_{i,j,k}}{2h_{x+}} + k(h_{y+} + h_{y-})h_{z-}\frac{T_{i-1,j,k} - T_{i,j,k}}{2h_{x-}} + \\
&k(h_{x+} + h_{x-})h_{z-}\frac{T_{i,j+1,k} - T_{i,j,k}}{2h_{y+}} + k(h_{x+} + h_{x-})h_{z-}\frac{T_{i,j-1,k} - T_{i,j,k}}{2h_{y-}} + \\
&k(h_{x+} + h_{x-})(h_{y+} + h_{y-})\frac{T_{i,j,k-1} - T_{i,j,k}}{2h_{z-}} + \\
&h^e(T_a - T_{i,j,k})(h_{x+} + h_{x-})(h_{y+} + h_{y-}) = 0 \qquad (4.31)
\end{aligned}
$$

where h^e is the effective heat transfer coefficient defined in Eq. (4.10). Using the analogy in Fig. 4.11, the thermal conductances G_1, G_2 and G_3 in Fig. 4.14(a) are found as

$$
G_1 = \frac{kh_{z-}(h_{y+} + h_{y-})}{2h_{x+}} \qquad (4.32)
$$

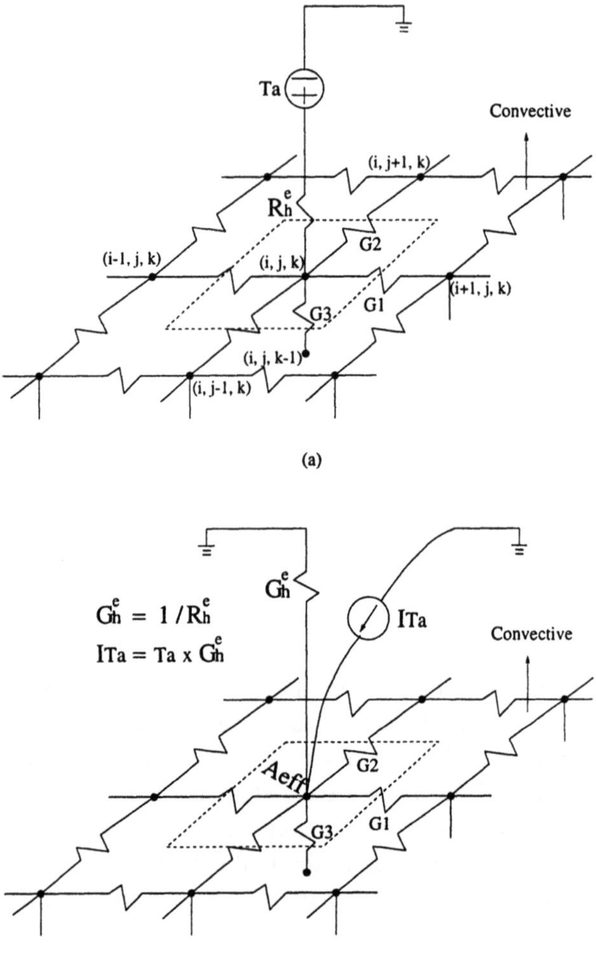

Figure 4.14. Equivalent thermal circuit at the convective boundary.

$$G_2 = \frac{kh_{z-}(h_{x+} + h_{x-})}{2h_{y+}} \tag{4.33}$$

$$G_3 = \frac{k(h_{x+} + h_{x-})(h_{y+} + h_{y-})}{2h_{z-}}. \tag{4.34}$$

Defining $A_{eff} = (h_{x+} + h_{x-})(h_{y+} + h_{y-})$ as the effective area of the grid (i, j, k), the thermal resistance associated with the convective heat transfer (R_h^e in Fig. 4.14(a)) is

$$R_h^e = \frac{1}{h^e A_{eff}}. \tag{4.35}$$

This boundary value problem can be solved as follows. First, the circuit in Fig. 4.14(a) is transformed to the equivalent circuit in Fig. 4.14(b), and a 3-D network containing only resistive elements and independent current sources is obtained (i.e., capacitive elements are open-circuited in the steady state). Next, a nodal analysis of this network is performed, and the system admittance matrix is constructed and solved. For the boundary condition other than the convective condition, a similar procedure follows by replacing h^e in Eq. (4.35) with ∞ for the isothermal condition, or with 0 for the insulated condition.

4.3.3 ANALYTICAL APPROACH

If both the heat diffusion equation Eq. (4.4) and the boundary condition Eq. (4.5) are *homogeneous*, the heat conduction problem can be conveniently solved by using the method of *separation of variables*. For *nonhomogeneous* problems, a general method such as the multiple *integral transform* (i.e., Fourier transform) and the associated multiple inversion technique [13] is required. The integral transform technique removes the space variables from the partial differential equation. It treats all space variables in the same manner with no inversion difficulties because all the integral transform and the inversion formula are well defined at the onset of the problem.

Consider a general 3-D problem in the finite ranges of $0 \le x \le a, 0 \le y \le b$ and $0 \le z \le c$, where a, b, c are the dimensions of a solid. The triple-integral and inversion formulae are defined as

$$\text{(Triple-Integral)} \quad \overline{T}(\beta_m, \nu_n, \eta_p, t) = \int_0^a \int_0^b \int_0^c K(\beta_m, x')K(\nu_n, y')K(\eta_p, z')T(x', y', z', t)dx'dy'dz' \quad (4.36)$$

$$\text{(Triple-Inversion)} \quad T(x, y, z, t) = \sum_{m=0}^{\infty}\sum_{n=0}^{\infty}\sum_{p=0}^{\infty} K(\beta_m, x)K(\nu_n, y)K(\eta_p, z)\overline{T}(\beta_m, \nu_n, \eta_p, t) \quad (4.37)$$

where $K(\beta_m, x)$, $K(\nu_n, y)$, $K(\eta_p, z)$ are the eigenfunctions and β_m, ν_n, η_p are the eigenvalues. The eigenfunctions and eigenvalues can be derived by simply solving the auxiliary homogeneous problem called the *Sturm-Liouville* problem [14]. The integral transform of the heat equation Eq. (4.4) can be taken by applying Eq. (4.36):

$$\int_0^a \int_0^b \int_0^c K(\beta_m, x)K(\nu_n, y)K(\eta_p, z)\left(\frac{\partial^2 T}{\partial x^2} + \frac{\partial^2 T}{\partial y^2} + \frac{\partial^2 T}{\partial z^2}\right)dxdydz \; + \frac{1}{k}\overline{g}(\beta_m, \nu_n, \eta_p, t) = \frac{1}{\alpha}\frac{d\overline{T}(\beta_m, \nu_n, \eta_p, t)}{dt} \quad (4.38)$$

where $\overline{g}(\beta_m, \nu_n, \eta_p, t)$ is the integral transform of $g(x, y, z, t)$. By using the Green's theorem, Eq. (4.38) can be transformed into the following expression which takes the boundary conditions into account:

$$\frac{d\overline{T}(\beta_m, \nu_n, \eta_p, t)}{dt} + \alpha(\beta_m^2 + \nu_n^2 + \eta_p^2) \cdot \overline{T}(\beta_m, \nu_n, \eta_p, t) = A(\beta_m, \nu_n, \eta_p, t) \qquad (4.39)$$

where

$$\begin{aligned} A(\beta_m, \nu_n, \eta_p, t) &= \frac{\alpha}{k}\overline{g}(\beta_m, \nu_n, \eta_p, t) + \\ &\frac{\alpha}{k}\Bigg[K(\beta_m, x)|_{x=0} \int_0^b \int_0^c f_1(y, z, t) K(\nu_n, y)\, K(\eta_p, z) dy dz + \\ &K(\beta_m, x)|_{x=a} \int_0^b \int_0^c f_2(y, z, t) K(\nu_n, y) K(\eta_p, z) dy dz + \\ &K(\nu_n, y)|_{y=0} \int_0^a \int_0^c f_3(x, z, t) K(\beta_m, x) K(\eta_p, z) dx dz + \\ &K(\nu_n, y)|_{y=b} \int_0^a \int_0^c f_4(x, z, t) K(\beta_m, x) K(\eta_p, z) dx dz + \\ &K(\eta_p, z)|_{z=0} \int_0^a \int_0^b f_5(x, y, t) K(\beta_m, x) K(\nu_n, y) dx dy + \\ &K(\eta_p, z)|_{z=c} \int_0^a \int_0^b f_6(x, y, t) K(\beta_m, x) K(\nu_n, y) dx dy\Bigg] \end{aligned} \qquad (4.40)$$

where f_1 - f_6 correspond to $f_i(x, y, z, t)$ in Eq. (4.5) for the six sides of the solid. The solution of the first-order differential equation Eq. (4.39) subject to the transformed initial condition

$$\overline{T}(\beta_m, \nu_n, \eta_p, t = 0) = \overline{F}(\beta_m, \nu_n, \eta_p) \qquad (4.41)$$

is straightforward and can be found in most differential equation texts [14]. Once $\overline{T}(\beta_m, \nu_n, \eta_p, t)$ is solved, $T(x, y, z, t)$ can be found by using the inversion formula Eq. (4.37).

For the steady-state case, Eq. (4.39) reduces to

$$k(\beta_m^2 + \nu_n^2 + \eta_p^2) \cdot \overline{T}(\beta_m, \nu_n, \eta_p) = A(\beta_m, \nu_n, \eta_p) \qquad (4.42)$$

where

$$\begin{aligned} A(\beta_m, \nu_n, \eta_p) &= \overline{g}(\beta_m, \nu_n, \eta_p) + \\ &K(\beta_m, x)|_{x=0} \int_0^b \int_0^c f_1(y, z) K(\nu_n, y) K(\eta_p, z) dy dz + \end{aligned}$$

$$
\begin{aligned}
&K(\beta_m, x)|_{x=a} \int_0^b \int_0^c f_2(y,z) K(\nu_n, y) K(\eta_p, z) dy dz + \\
&K(\nu_n, y)|_{y=0} \int_0^a \int_0^c f_3(x,z) K(\beta_m, x) K(\eta_p, z) dx dz + \\
&K(\nu_n, y)|_{y=b} \int_0^a \int_0^c f_4(x,z) K(\beta_m, x) K(\eta_p, z) dx dz + \\
&K(\eta_p, z)|_{z=0} \int_0^a \int_0^b f_5(x,y) K(\beta_m, x) K(\nu_n, y) dx dy + \\
&K(\eta_p, z)|_{z=c} \int_0^a \int_0^b f_6(x,y) K(\beta_m, x) K(\nu_n, y) dx dy. \qquad (4.43)
\end{aligned}
$$

Similar to the transient case, once $\overline{\overline{T}}(\beta_m, \nu_n, \eta_p)$ is solved, $T(x, y, z)$ can be found subsequently.

As an example, now we present the solution of the steady-state case where all four sides and the top surface of the chip are insulated, while the bottom surface is convective. In this case, the eigenfunctions are

$$
\begin{aligned}
K(\beta_m, x) &= \sqrt{2/a}\cos(\beta_m x) \quad \text{with} \quad \sin(\beta_m a) = 0 \\
K(\nu_n, y) &= \sqrt{2/b}\cos(\nu_n y) \quad \text{with} \quad \sin(\nu_n b) = 0 \\
K(\eta_p, z) &= \sqrt{2}\left[\frac{\eta_p^2 + H^2}{c(\eta_p^2 + H^2) + H}\right]^{1/2} \cos(\eta_p(c - z)) \\
&= Q_p \cos(\eta_p(c - z)) \;\text{ with }\; \eta_p \tan(\eta_p c) = H \;\text{ and }\; H = h_z^e/k
\end{aligned}
$$

where h_z^e is the effective heat transfer coefficient of the bottom surface. Note that when β_m in $K(\beta_m, x)$ is zero, the coefficient $\sqrt{2/a}$ of $K(\beta_m, x)$ has to be replaced by $\sqrt{1/a}$ in order to retain the eigenfunction normalities. Same argument also applies to ν_n in $K(\nu_n, y)$. Now $A(\beta_m, \nu_n, \eta_p)$ in Eq. (4.43) becomes

$$
\begin{aligned}
A(\beta_m, \nu_n, \eta_p) &= \overline{g}(\beta_m, \nu_n, \eta_p) + \\
&\quad K(\eta_p, z)|_{z=0} \int_0^a \int_0^b (h_z^e T_a) K(\beta_m, x) K(\nu_n, y) dx dy. \qquad (4.44)
\end{aligned}
$$

In Eq. (4.44), $\overline{g}(\beta_m, \nu_n, \eta_p)$ can be expressed as

$$
\begin{aligned}
\overline{g}(\beta_m, \nu_n, \eta_p) &= \frac{16Q_p}{\sqrt{ab}} \frac{1}{\beta_m \nu_n \eta_p} \sum_{i=1}^{n_r} g_i \cos(\beta_m x_{ic}) \sin(\frac{\beta_m}{2} x_{id}) \cdot \\
&\quad \cos(\nu_n y_{ic}) \sin(\frac{\nu_n}{2} y_{id}) \cos(\eta_p(c - z_{ic})) \sin(\frac{\eta_p}{2} z_{id}) \qquad (4.45)
\end{aligned}
$$

if $m \neq 0$ and $n \neq 0$, where (x_{ic}, y_{ic}, z_{ic}), (x_{id}, y_{id}, z_{id}) and g_i are the center coordinates, dimensions, and the power density of heat source i, respectively, and n_r is the number of heat sources. If $m = 0$ and/or $n = 0$, a similar expression for $\overline{g}$ can be derived. Note that the number of terms used in the infinite series expansion in Eq. (4.37) is actually finite in practical implementation. For instance, when the additional term does not change the temperature value by more than a small specified amount (e.g., 0.01 °C) or percentage, the series summation is terminated.

4.3.4 DISCUSSION

So far the above three different thermal analysis methods have been introduced. In general, the numerical method is more powerful and can be more easily applied to various kinds of heat conduction problems (e.g., linear, nonlinear, homogeneous, nonhomogeneous). The analytical method, on the other hand, is more restrictive in solving complex problems. For instance, if the thermal conductivity k is temperature dependent, some variable transformation technique such as the *Boltzmann's transformation* or the *Kirchhoff's transformation* [15] may need to be used in order to simplify the original problem to an analytically tractable level. Moreover, for systems containing composite material or multiple layers, special treatment is necessary in deriving the analytical solutions.

From the viewpoint of computational efficiency, the nonclosed form of the analytical triple series summation in Eq. (4.37) is more expensive than the numerical method [2], and this is aggravated when the number of points for which the temperature needs to be calculated is large (e.g., full-chip temperature profile estimation). For instance, the chip shown in Fig. 4.8 was simulated by both methods, and the temperatures of 400 mesh points on the chip were calculated. The numerical method required 26.09 seconds and the analytical one required 448 seconds of CPU time on SUN SPARC 10. The analytical method, however, has a clear advantage. Because it provides an explicit expression for the temperature of a point (x,y,z), it is very useful if one only needs to calculate the temperatures of some specified points (e.g., hot spots identified by the FTA method) rather than solving the temperature profile of the whole chip.

The FTA method can efficiently identify the hot spots. To demonstrate its speed advantage, both the FTA method and the numerical finite-difference method are used to identify the hot spot among the heat sources in Fig. 4.8. The speedup factor of FTA over the numerical method with increasing number of heat sources is plotted in Fig. 4.15. In the numerical simulation, one grid line is assigned for each heat source in both x and y directions.

The mixed use of the above thermal simulation methods are commonly seen for solving the temperature of complex systems. For instance, the FTA method was used in conjunction with the numerical method for mesh generation [3]. Areas with larger temperature gradient often require more grids for better accuracy. The FTA routine can be incorporated into the numerical thermal

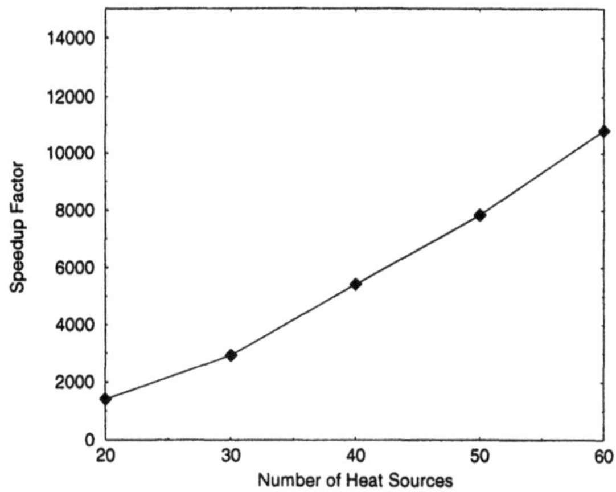

Figure 4.15. Speedup of FTA over numerical method.

simulator as a preprocessor to estimate the temperature gradient, which helps to determine a better initial guess of the grid distribution and therefore facilitate a more accurate and efficient thermal simulation. Moreover, in [16], a semi-analytical method to predict the printed circuit board package temperatures was proposed. In this approach, the analytical solution is derived to represent the temperature of each package layer, while the numerical iteration is needed to take into account the heat flow between layers and the final package temperature. In [17], a combination of the numerical and analytical methods to solve the temperature of the multilayered structure was also developed.

4.4. PACKAGE SIMULATION

Section 2 in the chapter presents the basic concept of the package modeling: The package related thermal resistances are first found, and then later incorporated into the 3-D bulk simulation in order to find the temperature distribution of the bulk. In the following, the packaging effects, the heat-flow paths through the package, and the calculation of the thermal resistance, will be explained in more detail.

4.4.1 MODELING OF THE CONVECTIVE BOUNDARIES

The effective heat transfer coefficient h^e in Eq. (4.10) describes how significantly heat transfers between the object and the ambient. Its value is determined by both the package structure and the efficiency of the heat removal process. Consider a solid with the dimensions of 1000 μm $\times$ 1000 μm $\times$ 250 μm. It

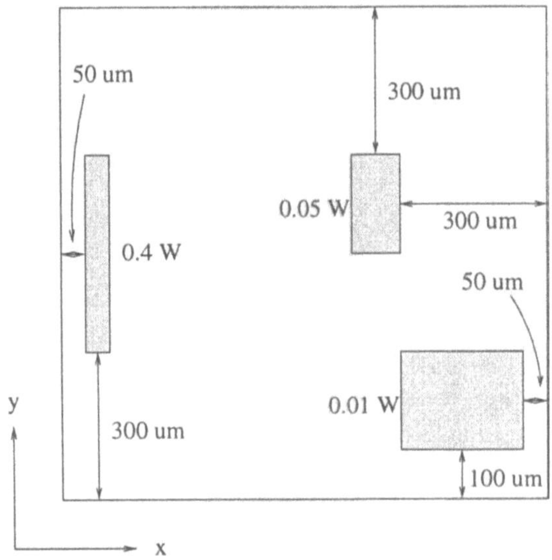

Figure 4.16. Layout of the solid containing three heat sources.

contains three heat sources with power values shown in Fig. 4.16. The heat transfer coefficient of the bottom sink is given to be 10,000 (W/m^2 °C).

Let us assume that the top and the four sides of the solid have the same h^e values of 0 (perfect insulation), 8.0 (natural convection), and 5,000. The simulated temperature profiles along the x direction at $y = 500$ μm are shown in Fig. 4.17 for these different h^e values. For a packaged solid under the natural convective condition, the boundaries of the top and all four sides are often approximated as perfectly insulated in the thermal simulation. It is verified by the results shown in Fig. 4.17, where the temperature difference under the two conditions is very small. However, if a solid is under a forced-convective condition (e.g., $h^e = 5,000$), its boundaries can no longer be modeled as perfectly insulated. As can be seen from Fig. 4.17, the forced convection greatly reduces the overall temperature. The packaging effect, therefore, plays an important role in temperature calculation and must be modeled correctly.

4.4.2 MODELING OF HEAT FLOW PATHS

Figure 4.18 shows a unit-level layout of a prototype ULSI high-performance chip in its initial design phase with the flip-chip packaging technology. Each unit contains several functional unit blocks (FUBs), and the estimated power values of all FUBs are given. There are in total more than three hundred FUBs in the chip. A cross-sectional view of the flip-chip package is shown in Fig. 4.19. The flip-chip bonding technology offers a better packaging solution, but also brings challenges for heat removal from the chip to the package (through the

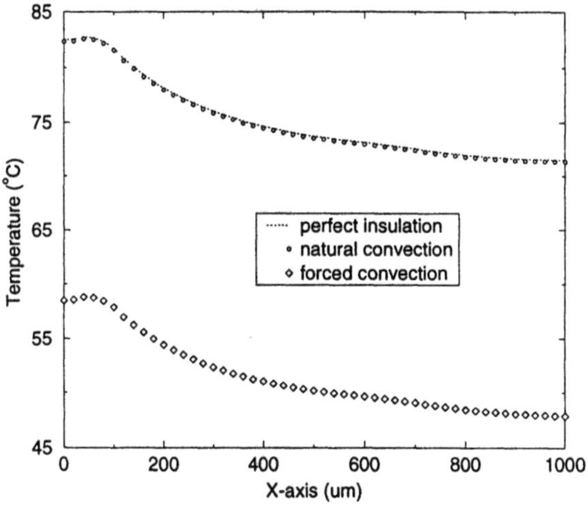

Figure 4.17. Temperature profiles along the x direction at $y = 500\ \mu$m for three different h^e values.

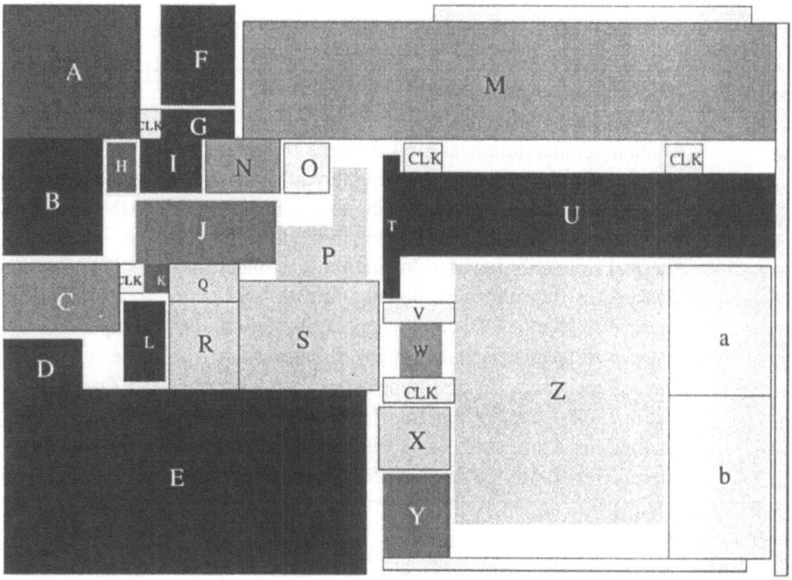

Figure 4.18. Unit-level layout of a high-performance chip.

bumps in Fig. 4.19). Furthermore, the heat sink (i.e., heat pipe) must be efficient enough to serve as the major heat removal path. From measurements, the temperature at the surface of the heat pipe is estimated to be 45 °C. The

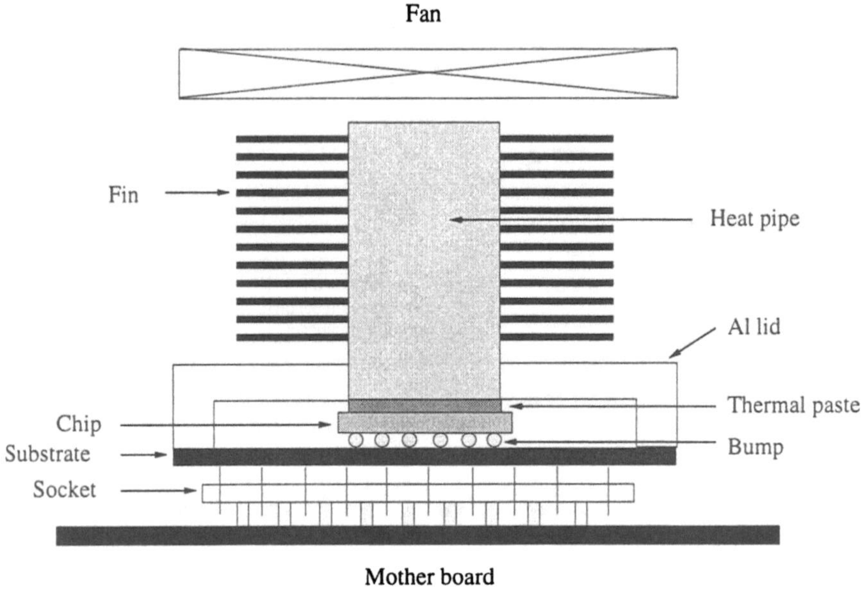

Figure 4.19. Cross-sectional view of a flip-chip package.

Table 4.4. Definition of the symbols in Fig. 4.20.

Ta_{27}	Ambient temperature (27°C)
Ta_{45}	Ambient temperature (45°C)
Ta_m	Ambient temperature near mother board
R_{car_down}	Thermal resistance for heat flowing through carrier down to the mother board
R_{car_side}	Thermal resistance for heat flowing through carrier aside to the lids
R_{bump}	Thermal resistance for heat flowing through bumps
R_{under}	Thermal resistance for heat flowing through underfills
R_{lid1}	Thermal resistance for heat flowing through lids to air
R_{lid2}	Thermal resistance for heat flowing through lids to the heat pipe
R_{tp}	Thermal resistance for heat flowing through the thermal paste
R_{h1}	Thermal resistance for heat transfer between lid surface and Ta_{27}
R_{h2}	Thermal resistance for heat transfer between lid surface and Ta_{45}
R_{hs}	Thermal resistance for heat transfer between the thermal paste and Ta_{45}
R_{hm}	Thermal resistance for heat transfer between the carrier and mother board

equivalent thermal circuit of Fig. 4.19 is shown in Fig. 4.20, and the symbol definitions are listed in Table 4.4.

To formulate the effective heat transfer macromodel, the thermal resistances of the carrier (R_{car_down}, R_{car_side}) and lids (R_{lid1}, R_{lid2}) among others need to be determined. To find the lumped value of R_{car_side}, for instance, the boundaries at four sides of the carrier is set to be in constant temperature

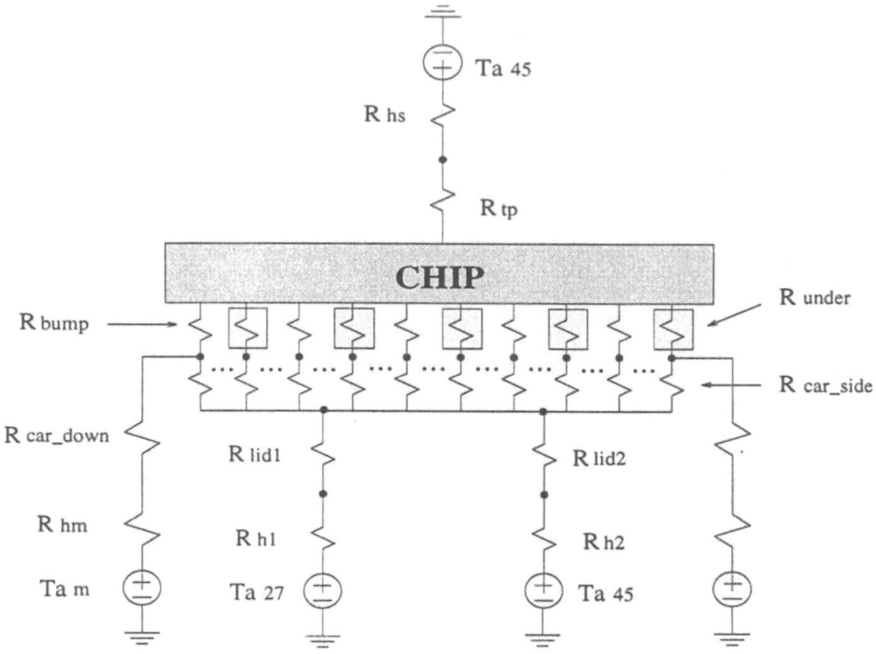

Figure 4.20. Equivalent thermal circuit of the flip-chip package.

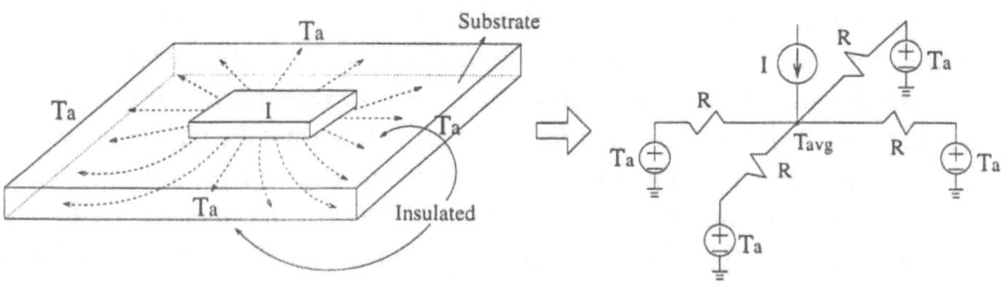

Figure 4.21. **Method to determine the thermal resistances for heat flowing through the carrier aside to the lids.**

T_a, and the top (except for the chip-carrier interface) and bottom surfaces of the carrier to be insulated. The lumped thermal resistance of the carrier can therefore be calculated as $R = (\frac{T_{avg}-T_a}{I/4})$, where I is the chip power, and T_{avg} is the average temperature at the chip-carrier interface which is precharacterized by performing numerical or analytical thermal simulation on the carrier. This procedure is graphically shown in Fig. 4.21. To determine the values of the thermal resistances accounting for the heat transfer between package surfaces

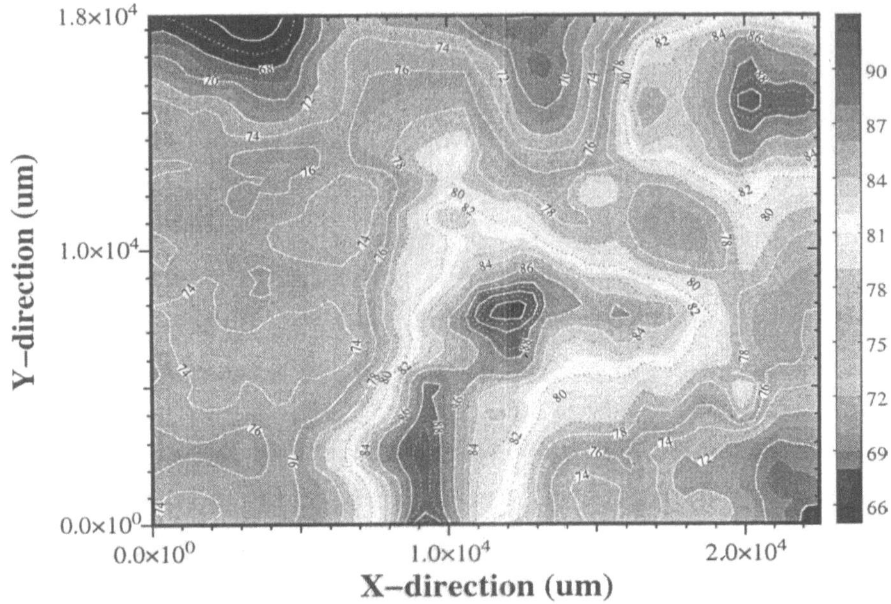

Figure 4.22. On-chip temperature contour for the first experiment.

and ambient (e.g., R_{h1}, R_{h2}, R_{hs}), formula $R = \frac{1}{h_p A}$ is used, where h_p is the heat transfer coefficient from package to ambient and A is the surface area.

Having discussed the package modeling, we now refer back to Fig. 4.20 and start the thermal simulation. In order to see how important the packaging effect is to the overall on-chip temperature, three different experiments are performed. In the first experiment, the contact resistance between the thermal paste and the heat pipe ($\approx$ 0.15 °C/W) is ignored. Figure. 4.22 shows a simulated temperature contour. In the second experiment, the contact resistance is taken into account, but it is assumed that no heat is flowing through the carrier (i.e., $R_{car\_side} = R_{car\_down} = \infty$). The simulation result is shown in Fig. 4.23. In the last experiment, both the contact resistance and the finite carrier thermal resistance are considered, and the result is shown in Fig. 4.24. It is clear that different packaging structures and materials have significant impact on the bulk temperature distribution. An electrothermal simulator can thus be used to guide the thermally reliable package design and improve the chip performance.

4.5. SUMMARY

This chapter deals with the topics related to the thermal simulation for VLSI systems.

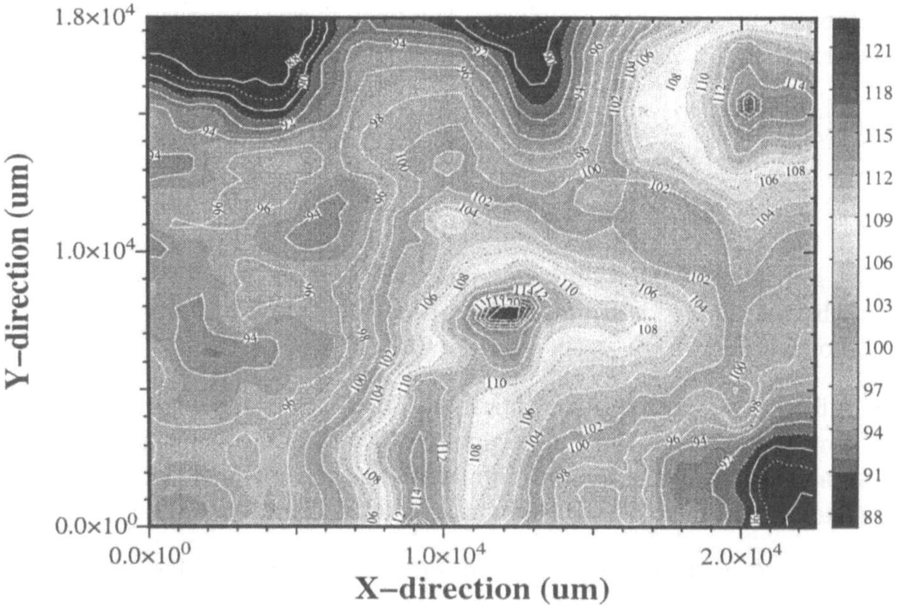

Figure 4.23. On-chip temperature contour for the second experiment.

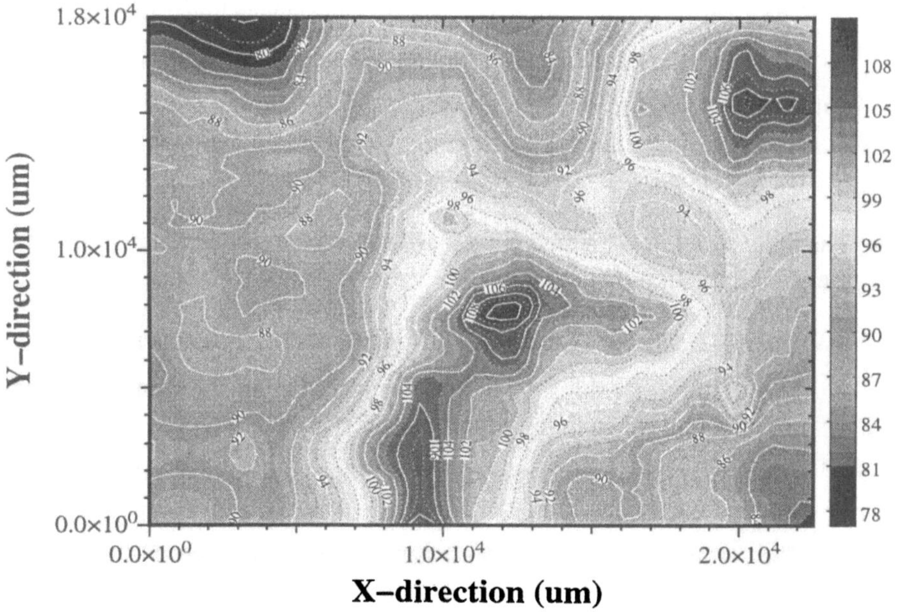

Figure 4.24. On-chip temperature contour for the third experiment.

- The heat diffusion equation for describing heat conduction in solids is provided. In addition, the thermal boundary conditions and the initial temperature condition are given. The three thermal boundary conditions are:
  - Isothermal (Dirichlet) condition
  - Insulated (Neumann) condition
  - Convective (Robin) condition
- A general description of the thermal simulation strategy for a simple chip substrate/package structure is given, and the effective heat transfer macro-model is described.
- A thermal simulation framework that is developed mainly for the VLSI systems is presented. This framework consists of fast thermal analysis (FTA), numerical/analytical thermal simulation, package thermal simulation, and interconnect thermal simulation (discussed in Chapter 6).
- The FTA method is developed for efficiently identifying the on-chip hot spots.
  - It is based on the formulation and approximation of the Green's functions.
  - The error function is approximated by a piecewise linear function
  - Six constraints are specified so that the original problem is simplified to a tractable level.
  - Experimental results show that the FTA method can quickly pinpoint the hot spots.
- The finite-difference method (FDM) is chosen in this book as the numerical thermal simulation method.
  - The basic concept of the finite-difference approach is provided.
  - The adaptive meshing technique for distributing the grid lines with variable grid space in the FDM method is introduced.
  - The analogy between the thermal circuit and electrical circuit is provided.
  - The formulae of thermal conductance and capacitance in the 3D grid system are derived. Similar expressions are derived for the system containing composite materials. Special care must be taken to model the boundary grids.

- The analytical thermal simulation method is presented.
    - It is based on the integral transform (Fourier transform) and the associated multiple inversion techniques
    - The derivation of the analytical formula for a general 3D problem is presented.
    - The eigenfunctions and the analytical solutions of the steady-state problem with special thermal boundary conditions are provided as an example.
- The above thermal simulation methods are compared. In general, the FTA method is the fastest, the numerical method is the most general, and the analytical method has the advantage when calculating the temperatures of a small number of spots.
- The package simulation method is presented. The importance of accurate package modeling is demonstrated. A prototype chip with the flip-chip package is used as a walk-through example:
    - Different heat flow paths in this structure are identified and modeled
    - The lumped thermal resistances of the packages are calculated
    - Three different experiments are performed. From the experimental results it can be seen that different packaging structures and materials can have significant impact on the bulk temperature distribution. An electrothermal simulator can be used to guide the thermally reliable package design.

References

[1] M. N. Ozisik, *Boundary Value Problems of Heat Conduction.* New York, NY: Dover, 1968.

[2] Y. K. Cheng, *Electrothermal Simulation and Temperature-sensitive Reliability Diagnosis for CMOS VLSI Circuits.* PhD thesis, Dept. of Electrical and Computer Engineering, University of Illinois at Urbana-Champaign, 1997.

[3] Y. K. Cheng and S. M. Kang, "An efficient method for hot-spot identification in ULSI circuits," in *Proceedings of the ACM/IEEE International Conference on Computer-Aided Design*, pp. 124–127, Nov. 1999.

[4] V. Dwyer, A. Franklin, and D. Campbell, "Thermal failure in semiconductor devices," *Solid-State Electronics*, vol. 33, pp. 553–560, May 1990.

[5] K. Fukahori and P. R. Gray, "Computer simulation of integrated circuits in the presence of electrothermal interaction," *IEEE Journal of Solid-State Circuits*, vol. 11, pp. 834–846, Dec. 1976.

[6] Y. K. Cheng, P. Raha., C. C. Teng, E. Rosenbaum, and S. M. Kang, "ILLIADS-T: An electrothermal timing simulator for temperature-sensitive reliability diagnosis of CMOS VLSI chips," *IEEE Transactions on Computer-Aided Design of Integrated Circuits and Systems*, vol. 17, pp. 668–681, Aug. 1998.

[7] G. Digele, S. Lindenkreuz, and E. Kasper, "Fully coupled dynamic electro-thermal simulation," *IEEE Transactions on VLSI Systems*, vol. 5, pp. 250–257, Sept. 1997.

[8] W. K. Chu and W. H. Kao, "A three-dimensional transient electrothermal simulation system for IC's," in *Proceedings of the THERMINIC Workshop*, pp. 201–207, Sept. 1995.

[9] V. A. M. N. Sabry, A. Bontemps and R. Vahrmann, "Realistic and efficient simulation of electro-thermal effects in VLSI circuits," *IEEE Transactions on VLSI Systems*, vol. 5, pp. 283–289, Sept. 1997.

[10] C. C. S. Wunsche and P. Schwarz, "Electro-thermal circuit simulation using simulator coupling," *IEEE Transactions on VLSI Systems*, vol. 5, pp. 277–282, Sept. 1997.

[11] B. D. J. P. Fradin, "Automatic computation of conductive conductances intervening in the thermal chain," in *Proceedings of the International Conference on Environmental Systems*, July 1995.

[12] J. F. Thompson, Z. Warsi, and C. W. Mastin, *Numerical Grid Generation*. New York, NY: North-Holland, 1985.

[13] J. W. Brown and R. V. Churchill, *Fourier Series and Boundary Value Problems*. New York, NY: McGraw-Hill, 1993.

[14] D. G. Zill, *Differential Equations with Boundary-Value Problems*. Prindle, Weber & Schmidt, 1986.

[15] W. F. Ames, *Nonlinear Partial Differential Equations in Engineering*. Academic Press, New York, 1965.

[16] J. N. Funk, M. P. Menguc, K. A. Tagavi, and C. J. Cremers, "A semi-analytical method to predict printed circuit board package temperatures,"

IEEE Transactions on Components, Hybrids, and Manufacturing Technology, vol. 15, pp. 675–684, Oct. 1992.

[17] V. Koval, I. W. Farmaga, A. J. Strojwas, and S. W. Director, "MONSTR: A complete thermal simulator of electronic systems," in *Proceedings of the ACM/IEEE Design Automation Conference*, pp. 570–575, June 1994.

Chapter 5

FAST-TIMING ELECTROTHERMAL SIMULATION

5.1. INTRODUCTION

The electrical simulator in the electrothermal simulation framework calculates the power and the timing of a circuit. It must generate accurate results so that the electrothermal behaviors are well predicted. It also needs to be computationally efficient in order to simulate VLSI circuits with acceptable run time.

Existing electrical simulators can be classified into the following classes: the circuit level, the timing level, the fast-timing level, the switch-level, and the logic level. The circuit-level simulators such as SPICE provide the most accurate voltage and current information for the circuit nodes. They are indispensable in modern digital and analog circuit design. However, they are computationally very expensive. The timing-level simulators [1, 2] are less costly than the circuit-level simulators but less accurate. In order to obtain orders of magnitude speedup over SPICE, techniques such as the table lookup and the event-driven simulation are employed in the timing-level simulators. The switch-level simulators [3, 4] model the MOS transistor as a switch controlled by the gate voltage. If the transistor is conducting, then the switch is closed; otherwise the switch is open. To provide the timing information at the switch level, simple delay models are often used.

The fast timing simulators [5, 6] were originally proposed to bridge the gap between the timing-level simulator and the switch-level simulator. They have the efficiency close to the switch-level simulators and the accuracy comparable to the timing-level simulators. The fast timing simulators generally possess the following features:

- the event-driven simulation is used
- the subcircuits are mapped into the macromodels or primitives

- the node equations are solved by the analytical solution.

The fast timing simulators are not only computationally efficient, but can also be built with the accurate device model for submicron technology. Therefore, they are capable of simulating modern VLSI circuits.

The electrothermal simulator, ILLIADS-T [7], was developed mainly for simulating the VLSI systems. It uses a fast timing simulator as its electrical simulation engine because of the advantages described above. ILLIADS-T was described in Chapter 1. In this chapter, it will be revisited with the focus on its simulation of a tester chip and other benchmark circuits. In the following, the underlying principles of the fast timing simulator used in ILLIADS-T will be described. The approach used to speed up the iteration process between the power and temperature calculation in ILLIADS-T will also be presented. The details of the tester chip design and the experiment setup for verifying the ILLIADS-T simulation result will be given. From the experimental results, it will be shown that the thermal effect can significantly impact the overall chip/circuit performance, even causing functional failures.

5.2. ILLIADS: A FAST TIMING SIMULATOR

The major shortcoming of typical fast-timing simulation approaches lies in the lack of simulation accuracy. A major source of inaccuracy is due to the use of overly simplified MOS transistor models for the submicron devices with various short-channel effects. The other source is due to the inadequate mapping of gates or subcircuits to the circuit macromodels. To address the first inaccuracy problem, the regionwise quadratic (RWQ) device model [9] was developed for the accurate device modeling and circuit simulation. The details of the temperature-dependent RWQ model were described in Chapter 3. To solve the second inaccuracy problem, a fast timing simulator, ILLIADS, was developed [8]. A new circuit primitive is used in ILLIADS to reduce the mapping error. The resulting node equations for the primitive are then solved for the node voltages. The overview of the primitive formation and solutions in ILLIADS will be given next, followed by its simulation strategy.

5.2.1 PRIMITIVE FORMATION AND SOLUTIONS

The generic circuit primitive of MOS digital circuits used in ILLIADS is shown in Fig. 5.1. This primitive contains multiple branches of NMOS and PMOS transistors, linear coupling capacitances $C(.)$, and linear conductances $g(.)$. The output node voltage and the loading capacitance are denoted as $V(\tau)$ and C_L, respectively. The applied terminal voltages are represented by $D(.)$ and $G(.)$.

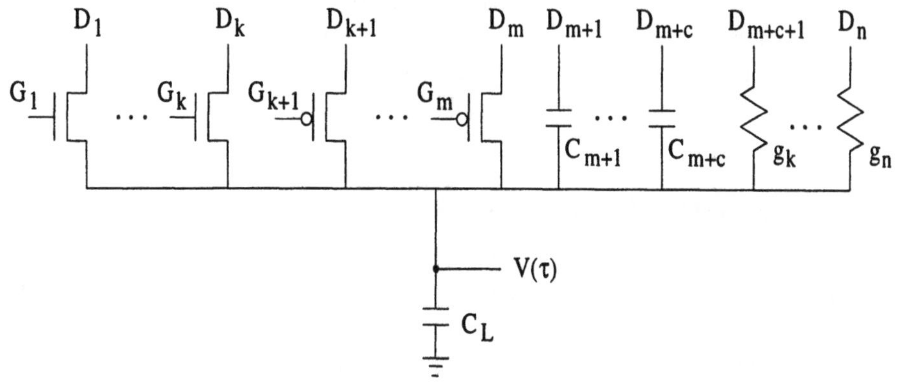

Figure 5.1. General MOS circuit primitive used in ILLIADS.

When the drain current i_k of MOS transistor k is modeled by a quadratic function of its terminal voltages, it can be expressed as

$$i_k = f((V - D_k)^p, (G_k - V - V_T)^q, (G_k - D_k - V_T)^r), \quad 1 \leq k \leq m \quad (5.1)$$

where p, q, r are integers that satisfy $0 \leq p, q, r \leq 2$, and V_T is the threshold voltage. The Shichman-Hodges model (see Eq. (3.1)) is a special case of Eq. (5.1). The current equations for the linear capacitors and conductors are given by

$$i_k = C_k \frac{d(D_k - V)}{d\tau}, \quad m < k \leq m + c \quad (5.2)$$

and

$$i_k = g_k(D_k - V), \quad m + c < k \leq n \quad (5.3)$$

respectively, where c is the number of capacitors in the primitive and n is the total number of parallel branches in the primitive. When the waveforms of the applied terminal voltages D_k and G_k are piecewise linearized, the state equation of the output node of the primitive in Fig. 5.1 can be written as

$$C_L \frac{dV}{d\tau} = \sum_{\text{MOS}} i_k + \sum_{\text{CAP}} C_k \frac{d}{d\tau}(D_k - V) + \sum_{\text{COND}} g_k(D_k - V) \quad (5.4)$$

with the initial condition $V(0) = V_0$. Substituting Eq. (5.1), Eq. (5.2) and Eq. (5.3) into Eq. (5.4), Eq. (5.4) can be recast to

$$\frac{dV}{d\tau} = kV^2 + (p_1\tau + p_0)V + (q_2\tau^2 + q_1\tau + q_0), \quad V(0) = V_0. \quad (5.5)$$

The coefficients k, p_1, p_0, q_2, q_1, and q_0 are written in terms of the MOS transistor model parameters, capacitances, conductances, and input signals.

Equation (5.5) belongs to the class of Riccati differential equations [11]. In the most general case, the analytical solutions for the Riccati differential equations can be found by using the hypergeometric functions [12]. An alternative approach is to use the power series method as done in ILLIADS. Detailed solutions of Eq. (5.5) and its degenerate forms are given in [10].

5.2.2 SIMULATION STRATEGIES

Given a circuit netlist, ILLIADS first partitions the circuit and groups the dc-connected blocks (DCCBs) using the breadth-first or depth-first search. A DCCB consists of a set of circuit nodes and elements that are connected through dc paths. Next, a directed graph (digraph) is constructed based on the connectivity of each DCCB. In order to detect the feedback loops, the vertices of the digraph are further partitioned into strongly connected components (SCCs). Each SCC contains DCCBs which can traverse to one another in the digraph. After the SCC partitioning, the circuit graph consists of *condensed* vertices which are either DCCBs or SCCs. The topological sort is then performed to obtain their temporal order. The SCC partitioning and the topological sort can be done simultaneously using the modified Tarjan's algorithm [13]. The procedures for SCC formation and topological sort are illustrated by the example in Fig. 5.2.

In ILLIADS, internal nodes (nodes that connect the dc paths of only one type of transistor, either NMOS or PMOS) are usually eliminated by merging serial or parallel transistors and forming the equivalent transistor. When necessary, however, internal nodes can also be simulated with the tradeoff of execution time. For serial merging of two transistors with transconductances β_1 and β_2, the transconductance of the equivalent transistor is given by $\beta_{eq} = \beta_1\beta_2/(\beta_1 + \beta_2)$. The equivalent gate signal is taken to be the weaker segment of the two gate signals (i.e., lower voltage for NMOS transistors and higher voltage for PMOS transistors). For parallel transistor merging, the equivalent transconductance is given by $\beta_{eq} = \beta_1 + \beta_2$, and the equivalent gate signal is the stronger segment of the two gate signals. Figure 5.3 illustrates the transistor merging and the internal node elimination process.

After internal node elimination, the remaining circuit is mapped into a primitive with the generic structure shown in Fig. 5.1. For example, the circuit after node elimination in Fig. 5.3 is mapped into the primitive given in Fig. 5.4. Next, the state equation is formed and the corresponding Riccati differential equation is solved analytically. The analogous output waveform can thus be obtained and used as the input to the next DCCB after piecewise linearization. For the circuit containing SCCs, the DCCBs inside the SCC are first ordered by a greedy algorithm where the DCCBs with the most external inputs and the least feedbacks are put in front of the queue. The waveform-relaxation method [14] is used to simulate the SCCs. The waveform-relaxation algorithm

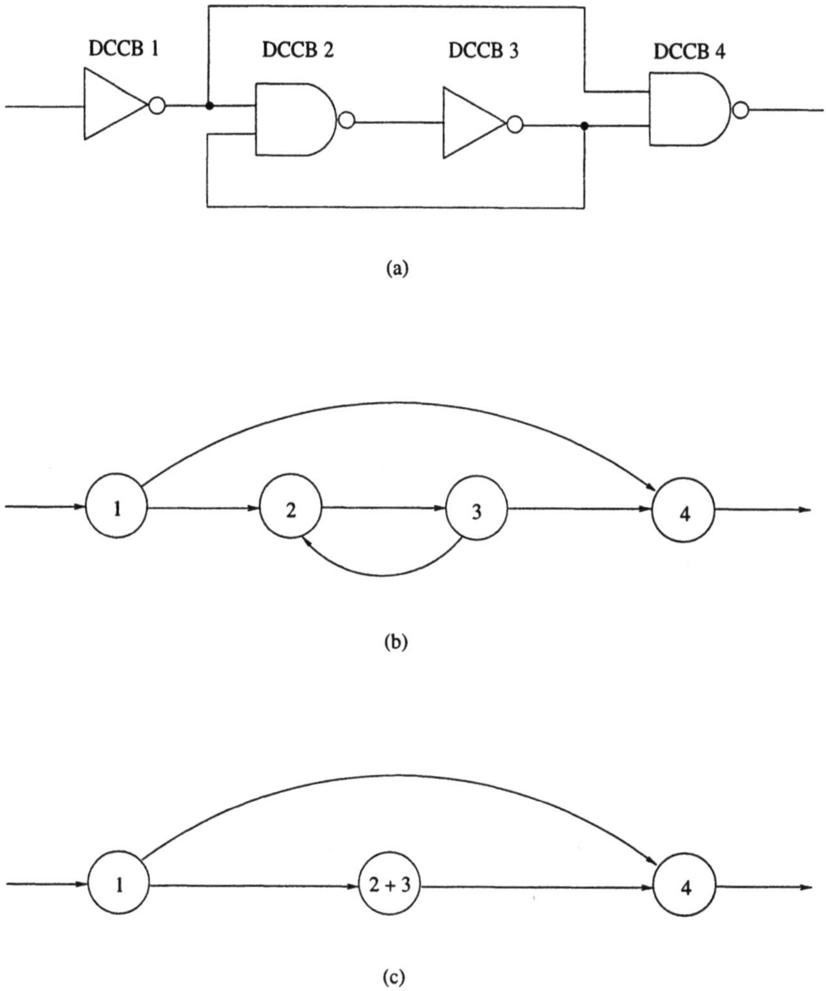

Figure 5.2. Illustrations of SCC formation and topological sort: (a) the original circuit, (b) the digraph representation, and (c) the condensed digraph after topological sort.

implemented in ILLIADS uses the partial waveform and time convergence and the dynamic windowing technique [8].

Note that the ILLIADS simulation efficiency is mostly contributed from the use of the analytical solutions of the Riccati differential equation. To formulate this equation, the drain current of the MOS transistor is assumed to follow the quadratic dependence on its terminal voltages. One such transistor model is the well-known Shichman-Hodges model. However, this model is not accurate enough for modeling the submicron devices. Therefore, a new quadratic-current model must be sought for in oder to improve the simulation accuracy

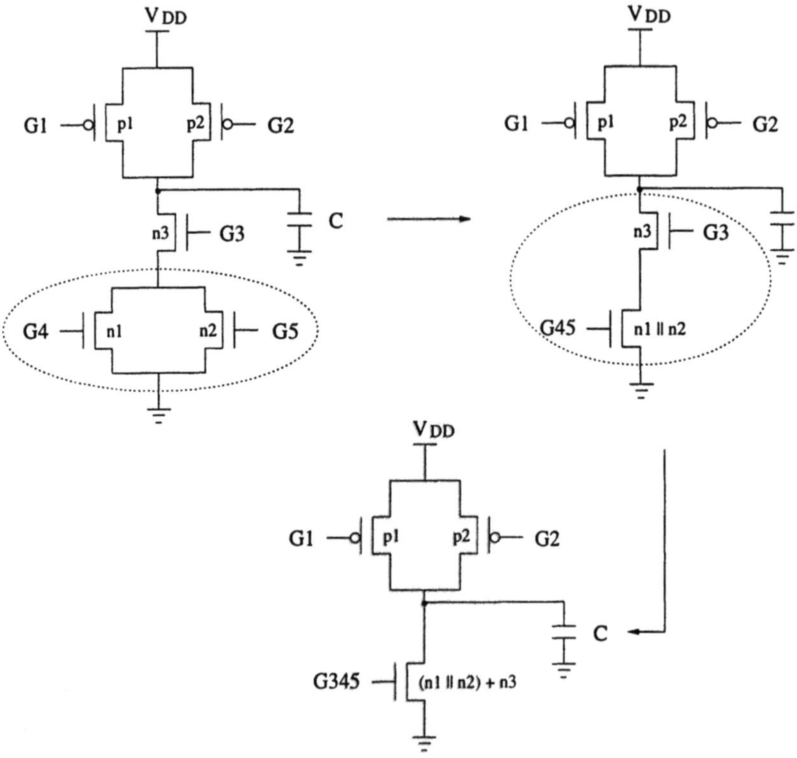

Figure 5.3. Example of transistor merging and internal node elimination.

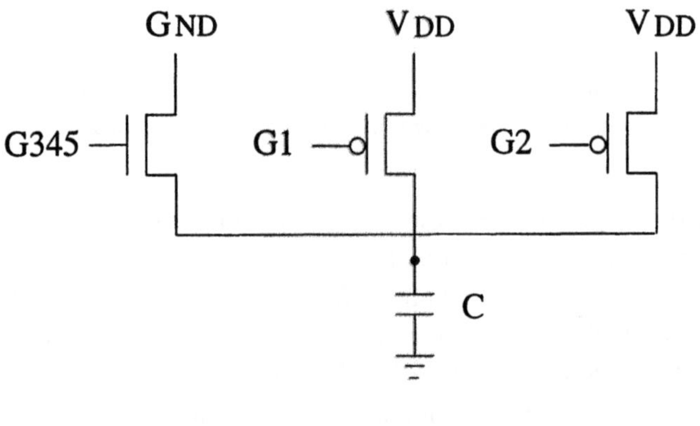

Figure 5.4. Primitive mapping for the circuit shown in Fig. 5.3 after the transistor merging process.

```
Algorithm DCCB POWER CALCULATION
  for (all DCCBs in the circuit)
    for (all simulated nodes in the DCCB)
      find voltage waveforms by solving the Riccati
      equation from the state equation of the node;
    for (all elements in DCCB connected to VDD)
      find the current waveform drawn from VDD;
    I_avg ← average the current waveform over the
            simulation period;
    P_avg ← I_avg × VDD;
End DCCB POWER CALCULATION
```

Figure 5.5. DCCB power calculation using ILLIADS.

in ILLIADS. The RWQ modeling technique [9] introduced in Chapter 3 was developed mainly for this purpose.

5.2.3 POWER ESTIMATION USING ILLIADS

Given the input patterns, the average power of each DCCB (or gate) in the circuit can be calculated by using ILLIADS. In the ILLIADS-T electrothermal simulation flow, each DCCB is treated as a heat source and its power value is input to the thermal simulator. The procedure for the power calculation in ILLIADS is shown in Fig. 5.5. It is similar to the power-meter method proposed in [15]. However, instead of building the power-meter circuitry, ILLIADS directly solves the Riccati differential equation for each DCCB primitive and finds the current waveform drawn from the power supply for all branches that are connected to it. Like the power-meter method, ILLIADS can accurately calculate both *dynamic* and *short-circuit* power.

5.3. INCREMENTAL ELECTROTHERMAL SIMULATION IN ILLIADS-T

The temperature-induced variations of the short-circuit power and the switching activity can affect the average power consumption of the logic gates. On the other hand, the changing power values can significantly perturb the original temperature distribution. As described in Chapter 1, because the power consumption and the temperature distribution are strongly dependent on each other, the decoupled electrothermal simulation flow in ILLIADS-T requires several iterations between the power and temperature calculation. When temperature and power converge to the values in the previous simulation run, the final steady-state solution is found.

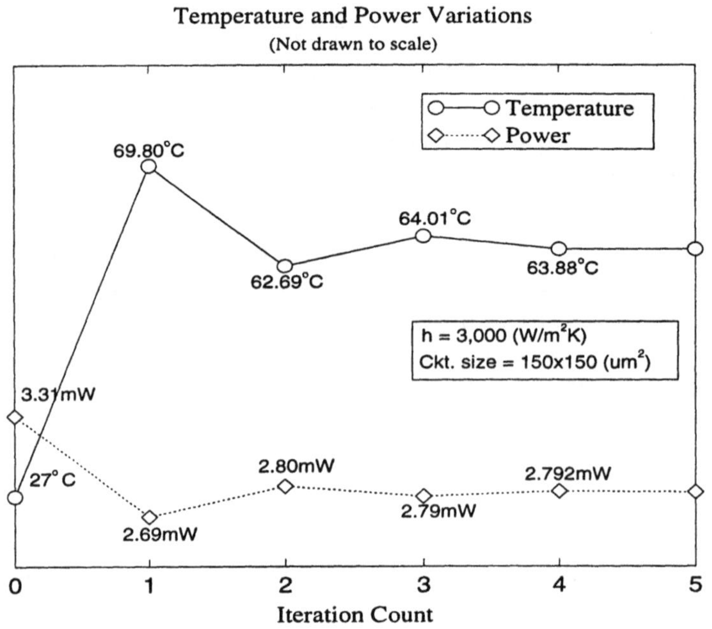

Figure 5.6. Convergence plot for power and temperature.

The above observation is shown in Fig. 5.6, where a nine-stage ring oscillator is simulated by ILLIADS-T. The power and temperature values are recorded during the iteration process. In Fig. 5.6, the temperature difference between two successive simulation runs becomes smaller as the number of iterations increases. This trend indicates the feasibility of *incremental simulation*[16], in which the circuit parameters vary by small amounts compared to their previous values.

Consider a circuit containing blocks that are ordered for simulation based on the fanin-fanout relationship. In ILLIADS-T, if the temperature difference in a block between current (perturbed) and previous (nominal) runs exceeds a prescribed threshold T_THRLD, then this block is considered to have local temperature variation and will be marked with T_VAR. If a block is marked with T_VAR, it is not considered to be incrementally *latent* and it needs to be simulated. The resimulated waveforms then serve as the nominal waveforms for the next simulation run. However, if a block is not marked with T_VAR, then the nominal and perturbed waveforms for all of the inputs to the block are compared. If the difference between them is less than a user-specified threshold, the block is marked as being incrementally latent and is not simulated (i.e., its perturbed solution will be the same as its nominal solution). On the other hand, if the difference in any of the inputs is larger than the threshold, the block is not considered to be latent and is simulated.

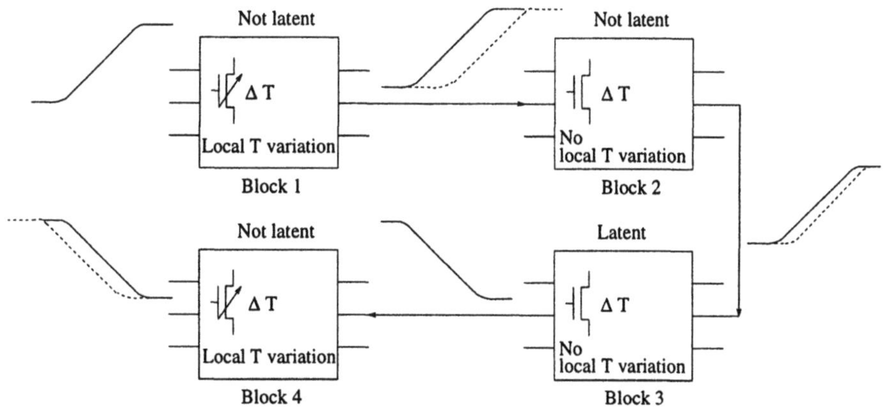

Figure 5.7. Illustration of incremental latency; the nominal waveforms are shown in solid lines, while the perturbed waveforms are in dashed lines.

This procedure is illustrated in Fig. 5.7, where Block 1 and Block 4 have local temperature variations and are incrementally resimulated. For Block 2, there is no local temperature variation, but the difference between its nominal and perturbed input signals is large and it, too, is resimulated. However, Block 3 is considered to be latent because besides having no temperature variation, the difference between its nominal and perturbed inputs is very small. Thus, for Block 3, the incremental simulation is skipped and its perturbed solution is assumed to be the same as its nominal results.

Note that ILLIADS-T identifies the blocks having local temperature variations *dynamically* in each new simulation run. In other words, a block can be marked with T_VAR in one run, but is not marked in another. As the number of iterations increases in the ILLIADS-T simulation, the advantage of the incremental approach will appear even greater because it is expected that a large number of latent cases will be detected. For larger circuits, it is also expected that the computational savings of latency will be more pronounced, because a larger number of blocks will be latent. Furthermore, for circuits with a larger temperature gradient (e.g., due to either a large power density variation or a special kind of boundary condition), the incremental technique will be even more effective. The simulation speedup due to the incremental approach will be presented later in this chapter.

5.4. TESTER CHIP DESIGN AND CALIBRATION

A tester chip was designed for the verification of ILLIADS-T simulation accuracy [7]. It was fabricated using 0.8 μm CMOS technology and packaged by MOSIS. Figure 5.8 shows the microphotograph of the chip, where the blocks I, III and V are high-frequency 3-stage ring oscillators designed in a standard

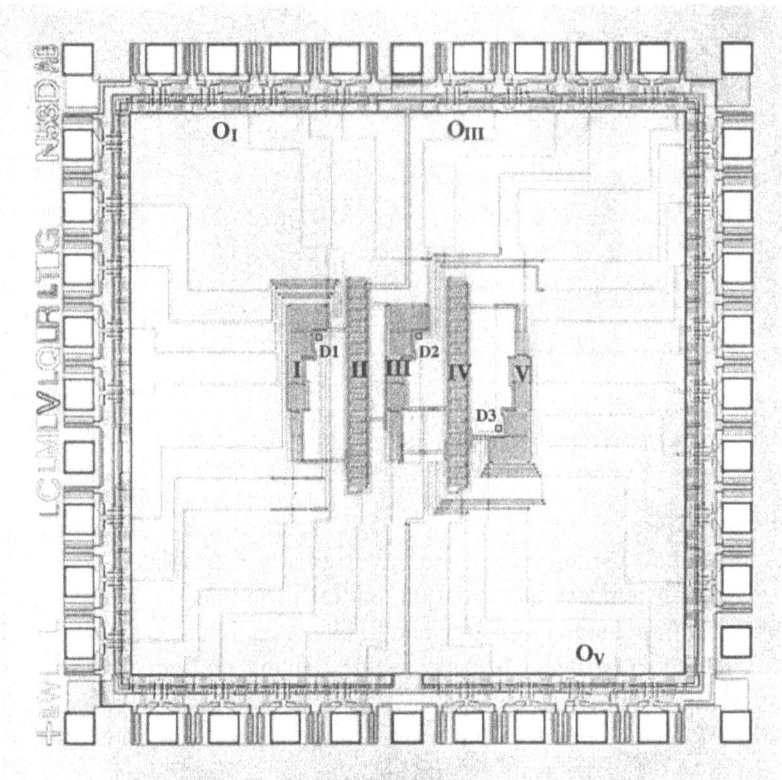

Figure 5.8. Microphotograph of the tester chip; long blocks are Rosc149s and short blocks are Rosc3s.

super-buffer configuration. Blocks II and IV are 149-stage ring oscillators, and the three small dots (D1, D2, D3) are temperature-monitoring diodes. Henceforth, the 3-stage and 149-stage oscillators are denoted as Rosc3s and Rosc149s, respectively. Each ring oscillator has an enable signal which is used to activate or deactivate the oscillator.

In this design, the power is mainly dissipated by the Rosc3s. The on-chip temperature can be determined by measuring the voltage drop across the forward-biased diodes according to $V_F = (kT/q)\ln(I_F/I_s(T) + 1)$, where I_F is the forward-bias current provided by a constant current source. The reason why the diode is measured in the forward-biased mode is that in the reverse-biased mode, in addition to the thermal generation of carriers, other generation-recombination mechanisms are also present to affect the accuracy of the measurement.

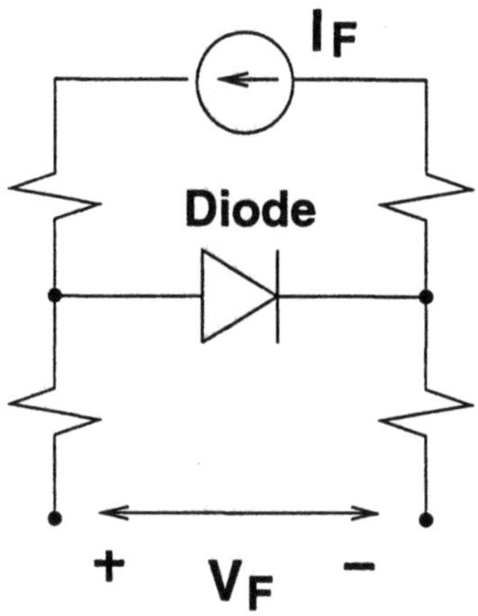

Figure 5.9. Four-terminal configuration for diode measurement.

Figure 5.9 shows the diode circuit designed for the temperature measurement. Because the voltage drop across the lead resistance of the diode is also a function of the temperature, a four-terminal configuration is used to cancel out the voltage drop in the test leads. The diodes were calibrated individually by measuring V_F at different temperatures. The diode temperature was controlled by placing the chip upside-down on a hot plate after the chip lid had been removed. The temperatures on the surface of the hot plate were accurately determined by placing a *thermistor* on the plate and measuring its resistance values. These values were then translated to the temperatures of the thermistor, namely, the temperatures of the hot plate.

The I_F values were forced small (15 - 20μA) to ensure that there is no self heating from the diode. An example of the calibration data is shown in Fig. 5.10, where both the stepwise I-V data and the glitch at 45 °C were caused by the measuring resolution. When the tester chip was operating, the local temperature near the diode was determined by using the calibration data. The package thermal parameters are also calibrated based on the MOSIS handbook for the DIP40 package. The effective heat transfer coefficient of the chip bottom (h^e in Eq. (4.10)) was determined to be 8,689 (W/(m^2 °C)) with all other sides insulated.

5.5. VERIFICATION OF ILLIADS-T

During the tester chip experiments, Rosc149s were always activated while the chip power consumption was varied by activating different Rosc3s. Depending on the on/off status of the Rosc3s, eight unique experiments can be performed as shown in Table 5.1. For instance, the ILLIADS-T-simulated

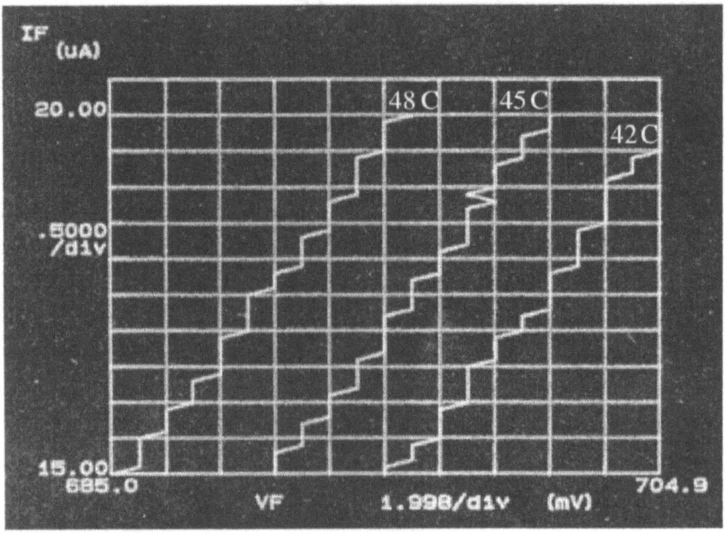

Figure 5.10. Diode calibration example (D1).

Table 5.1. Activation status of Rosc3s.

Expt. #	1	2	3	4	5	6	7	8
Block I III V	111	110	101	100	011	010	001	000

1: on 0: off.

temperature profiles for Expt. 2 (block I and III on, block V off) and Expt. 1 (all blocks on) are shown in Figs. 5.11 and 5.12, respectively. Note that the power dissipation from the output buffers of blocks I, III, and V (O_I, O_{III}, and O_V in Fig. 5.8) was also taken into account. The simulated and measured diode temperatures for all eight experiments are compared in Figs. 5.13 - 5.15. In these figures, the error bars show the spread of the measured data. Good agreement between measured and simulated temperatures was found.

ILLIADS-T was also used to predict the frequency shift of Rosc149s due to the local temperature rise. The mobility-temperature relationship was extracted from frequency measurements on block II, and the mobility model Eq. (3.19) was employed to obtain the optimized fitting parameters A_1 - A_4. Next, the mobility model was used in ILLIADS-T to predict the frequency shift of block IV for different cases. The results were compared with the measured data, and the measured and simulated waveforms of block IV are shown in Fig. 5.16 - Fig. 5.19.

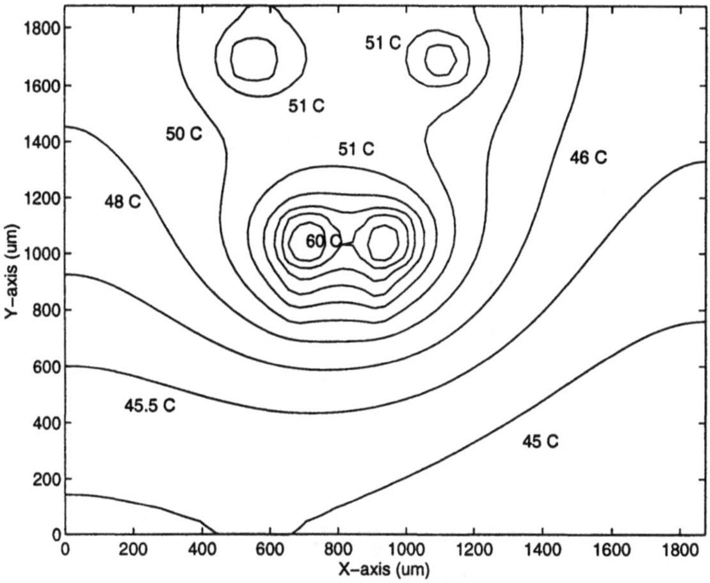

Figure 5.11. Simulated temperature profile for Expt. 2.

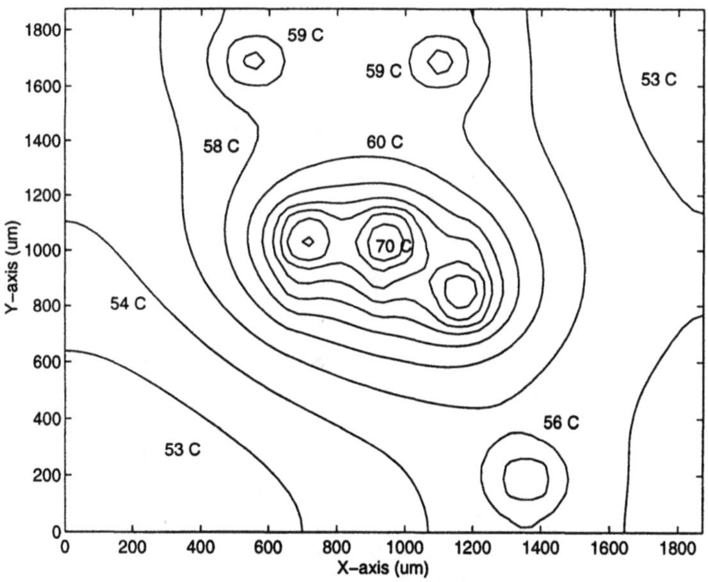

Figure 5.12. Simulated temperature profile for Expt. 1.

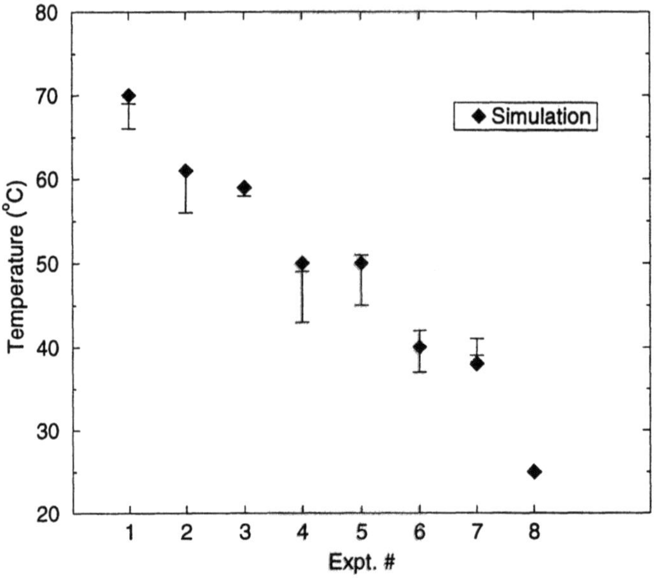

Figure 5.13. Comparison between simulated and measured temperatures for D1.

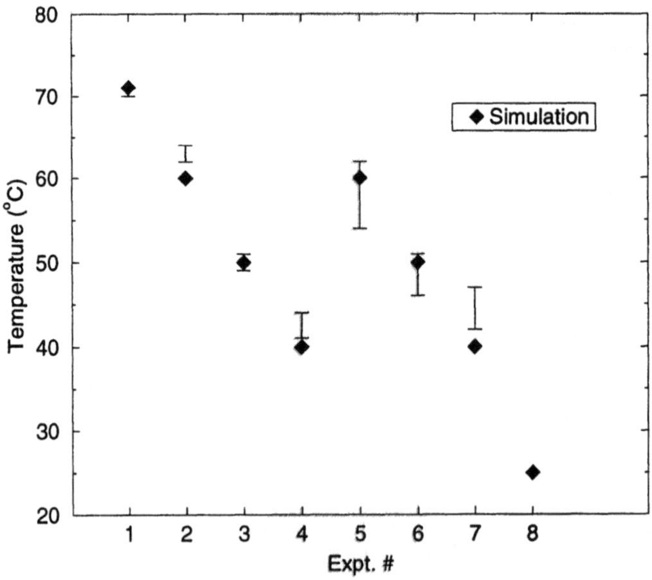

Figure 5.14. Comparison between simulated and measured temperatures for D2.

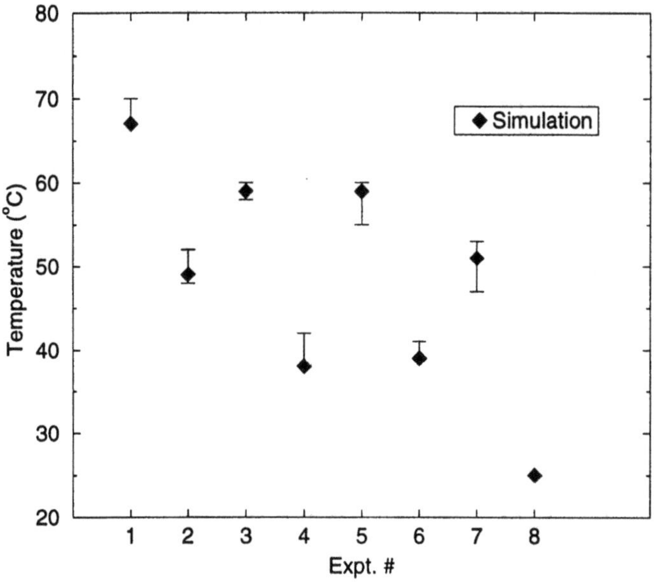

Figure 5.15. Comparison between simulated and measured temperatures for D3.

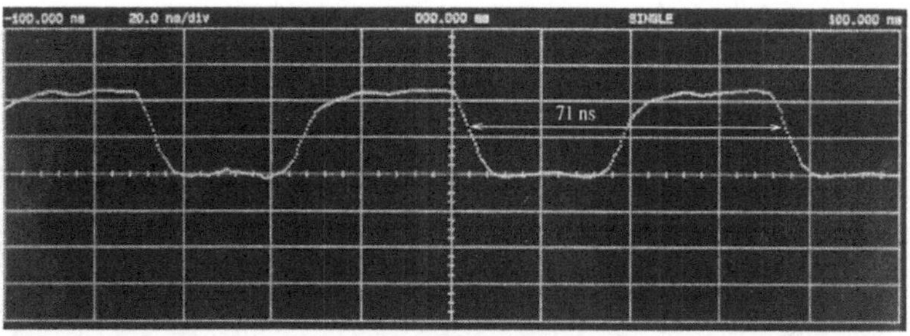

(a)

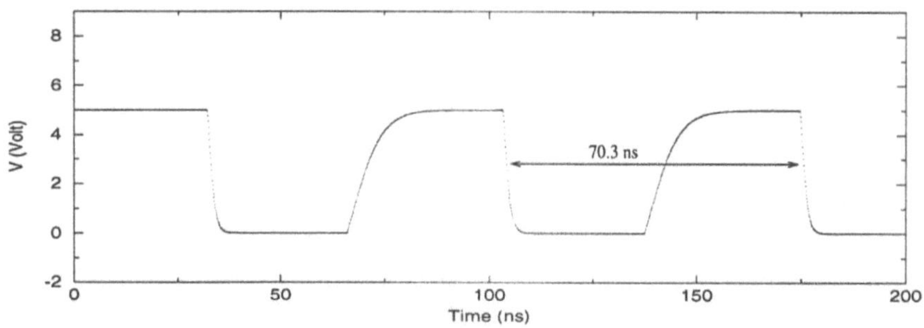

(b)

Figure 5.16. (a) Measured and (b) simulated waveforms for Expt. 8.

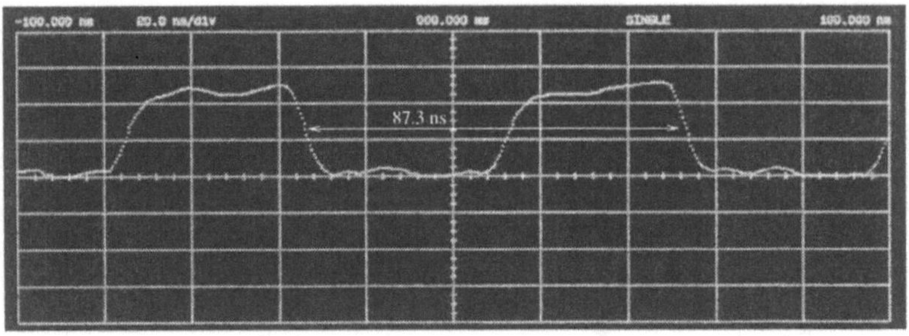

(a)

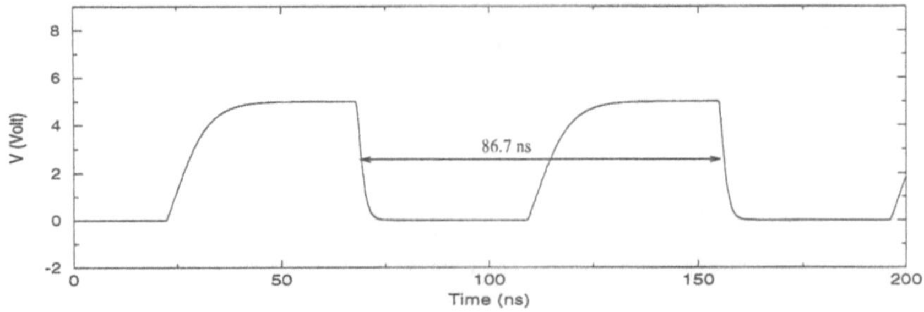

(b)

Figure 5.17. (a) Measured and (b) simulated waveforms for Expt. 7.

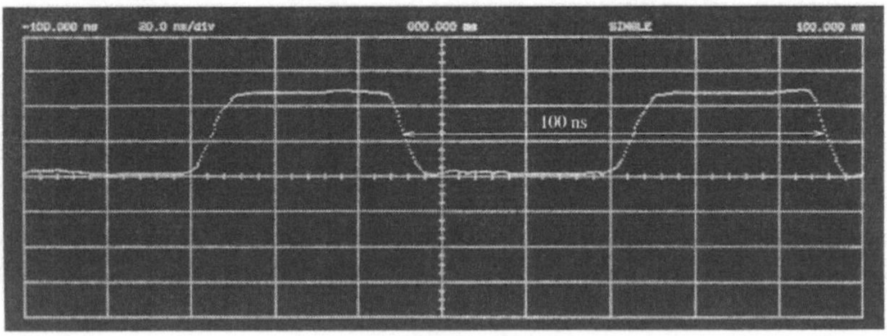

(a)

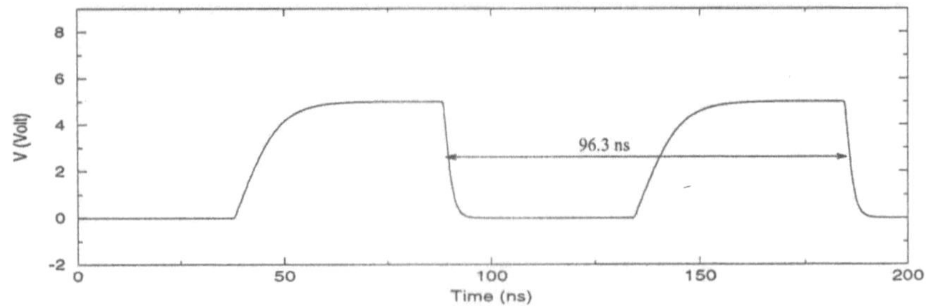

(b)

Figure 5.18. **(a) Measured and (b) simulated waveforms for Expt. 5.**

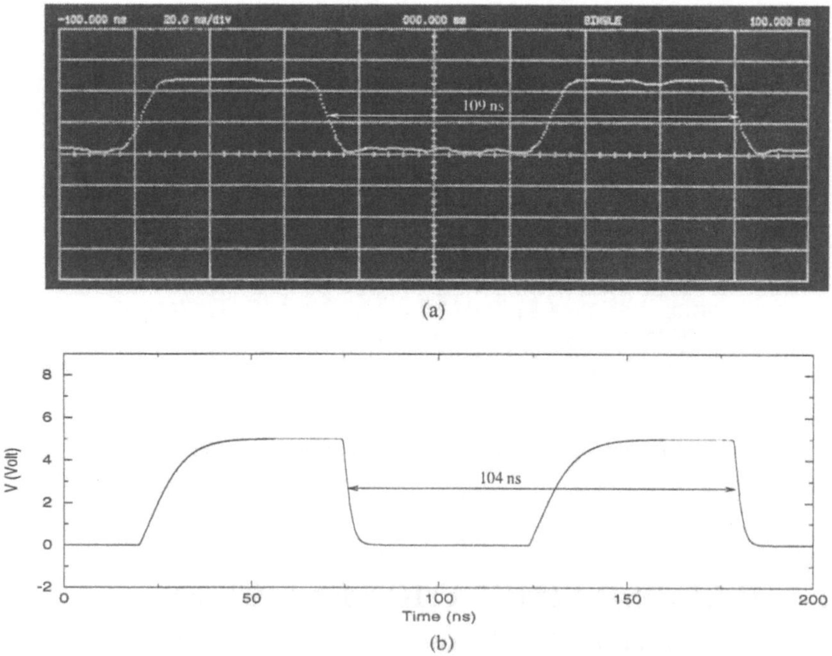

Figure 5.19. (a) Measured and (b) simulated waveforms for Expt. 1.

Table 5.2. ILLIADS-T simulation results of the tester chip.

Tester chip	P_{avg} [Watt]	T_{blk4} [°C]	Freq. shift [MHz]	CPU time[1] [sec]
Expt. 7	0.350	44.17	14.07 → 11.53	422
Expt. 5	0.636	56.03	14.07 → 10.38	650
Expt. 1	0.882	63.43	14.07 → 9.62	822

[1]On SUN SPARCstation 10 (numerical thermal simulation used).

Additional simulation results are presented in Table 5.2, where P_{avg} is the average power consumption of the chip (including output buffers), and T_{blk4} is the average temperature of block IV. The oscillation frequencies of block IV before and after electrothermal simulation are shown in the fourth column of Table 5.2. Note that as the temperature increases, the oscillation frequency is significantly lowered and, consequently, so is the power. Therefore, the power values listed in Table 5.2 were calculated at the simulated operating temperature. The total CPU time for the electrothermal simulation (i.e., numerical thermal simulation was used) is shown in the last column in Table 5.2.

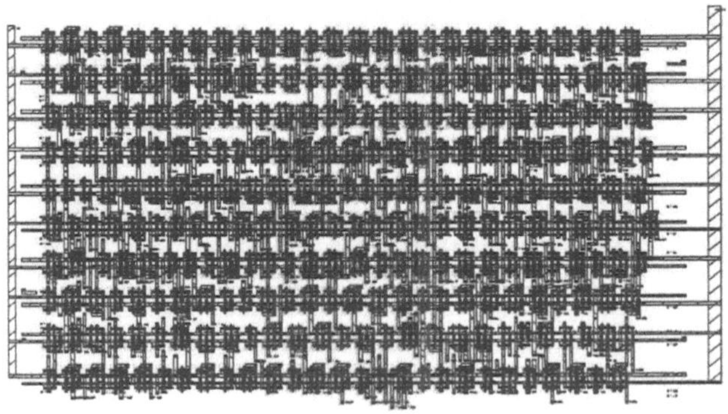

Figure 5.20. Layout of the 10-bit negative adder.

5.6. ILLIADS-T SIMULATION EXAMPLES

In the following, the ILLIADS-T simulation results for a number of circuits are demonstrated. The top and the four sides of the circuits are assumed insulated, while the effective heat transfer coefficient of the bottom surface is assumed to be 3,000 (W/(m^{2}K)) for all test circuits. A 10-bit negative adder is first considered with the layout shown in Fig. 5.20. The layout was generated by a synthesis tool iCGEN [17]. Simulation results are listed in Table 5.3, where n_{tran} and n_{hsrc} are the numbers of transistors and heat sources in the circuit, respectively, P_{avg} is the average total power consumed in the circuit, T_{avg} is the average circuit temperature, and n_{run} is the number of repeated simulation runs before convergence. The symbol S_{fac} denotes the speedup factor of the electrothermal timing simulation, which is computed as the ratio of the total (i.e., including all simulation runs) transient analysis time without the incremental simulation to the transient analysis time with the incremental simulation. Table 5.3 also presents the ILLIADS-T simulation results for several other circuits such as HIGHWAY [18], ALU and control, and a 16-bit multiplier.

Another simulation example is shown in Fig. 5.21. This chip contains two ISCAS85 benchmark circuits, C3540 and C6288, one negative adder, and two three-stage ring oscillators identical to Rosc3 in Fig. 5.8. The packaging structure for the chip is shown in Fig. 5.22 and the corresponding thermal parameters are given in Table 5.4. The heat transfer coefficient between the heat sink and the ambient is assumed to be 12,000 (W/(m^{2}K)). Simulation results are presented in Table 5.5. Here, T_{max} and T_{min} are the maximum and minimum temperatures of individual circuits; they were pinpointed by the

Table 5.3. ILLIADS-T simulation results.

Circuit	Ckt. size	n_{tran}	n_{hsrc}	Freq.	P^{1}_{avg}
Unit	μm^2	-	-	MHz	mW
10-bit Neg. Adder	450 × 300	868	216	100	12.07
HIGHWAY	150 × 210	248	17	33	1.92
ALU and Control	1730 × 1540	5842	1656	200	67.67
16-bit Multiplier	2180 × 2330	11016	3001	100	289.32

[1] Under steady-state temperature distribution.

Table 5.3 (continued). ILLIADS-T simulation results.

Circuit	T_{avg}	n^{2}_{run}	CPU time[3]	S_{fac}
Unit	°C	-	sec	-
10-bit Neg. Adder	49.40	3	31.35	1.24
HIGHWAY	46.15	3	23.01	1.21
ALU and Control	35.03	2	211.76	1.00
16-bit Multiplier	45.75	3	738.08	1.28

[2] Convergence criterion for iteration: $(\Delta T/T_{rise}) < \%1$.
[3] On SUN SPARCstation 10 (numerical thermal simulation used).

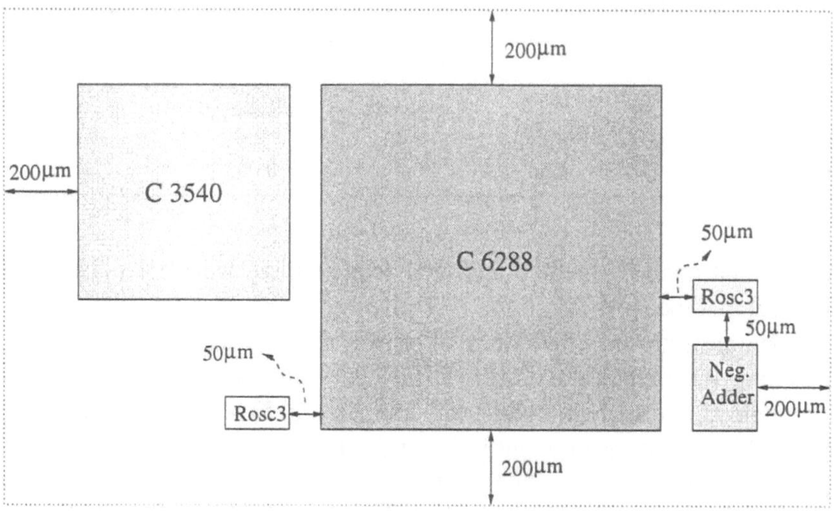

Figure 5.21. Layout of the simulated chip.

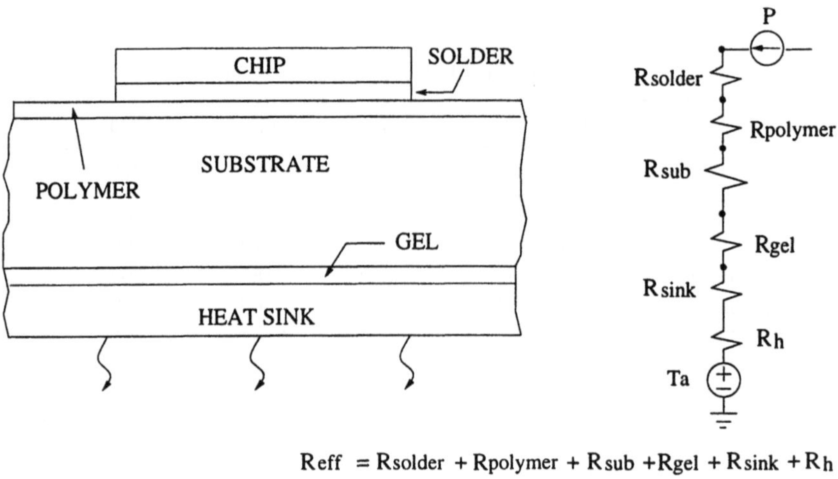

Figure 5.22. Packaging structure used in the simulation example.

Table 5.4. Packaging parameters for thermal simulation.

Parameter	Units	Bulk	Bump	Polymer
Material	-	doped Si	Sn/Pb	polymide
Thermal Conductivity	W/(mK)	98.40	53.4	0.25
Thickness	mm	0.25	0.025	0.005

Table 5.4 (continued). Packaging parameters for thermal simulation.

Parameter	Units	Substrate	Gel	Bottom Sink
Material	-	doped Si	silicone gel	Al
Thermal Conductivity	W/(mK)	98.40	0.4	216.5
Thickness	mm	0.5	0.005	1.0

FTA thermal analysis method and their temperatures were calculated by the analytical method.

To demonstrate the importance of performing the temperature-dependent simulation, the output waveforms at bit ten of the negative adder with and without electrothermal simulation are compared in Fig. 5.23. The output was generated by asserting an input pulse at the first bit with the pulse width 2.5 ns. A logic fault was identified via the electrothermal simulation. This fault was caused by the combination of temperature rise and unbalanced circuit design. It can be observed that the pulse width of the output waveform, even at room temperature, is narrower than the input pulse width (2.5 ns) due to the

Table 5.5. ILLIADS-T simulation results.

Circuit	Ckt. size	n_{tran}	n_{hsrc}	Freq.	T_{max}	T_{min}	n_{run}^{1}	CPU[2]
Unit	μm^2	-	-	MHz	°C	°C	-	sec
10-bit								
Neg. Adder	450 × 300	868	216	100	47.17	38.07	-	-
C3540	1730 × 1540	5842	1656	200	36.82	31.61	-	-
C6288	2180 × 2330	11016	3001	100	44.55	33.10	-	-
							3	3142

[1]Convergence criterion: $(\Delta T/T_{rise}) < \%1$.
[2]On SUN SPARCstation 10 (numerical thermal simulation used).

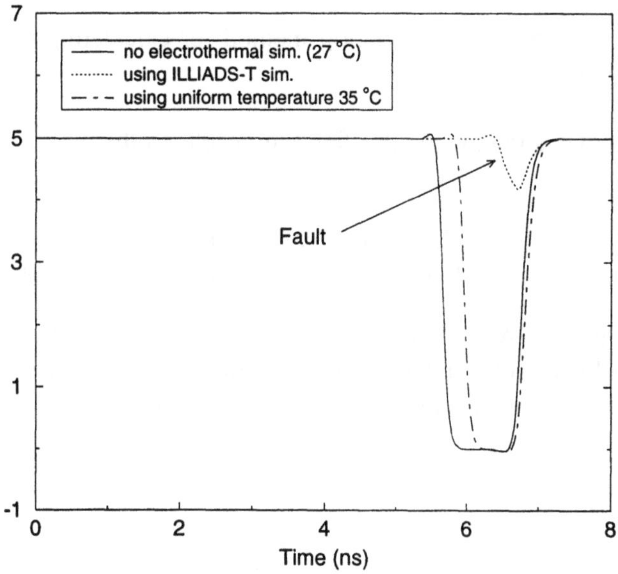

Figure 5.23. Output waveforms of the 10-bit negative adder.

unbalanced rise and fall path delays. This unbalance is aggregated at higher temperature and, finally at around 47 °C, the fault occurs before the output waveform can fully switch to ground. The average temperature of the whole layout in Fig. 5.21 was calculated to be 35 °C using ILLIADS-T. Note that when the temperature of the negative adder was assigned to 35 °C as would be used in conventional simulations, the logic fault still could not be detected as shown in Fig. 5.23. Simulation results suggest that the on-chip temperature variation must be considered in timing verification, and ILLIADS-T is one of the useful tools to ensure that the specified timing constraints are met.

5.7. SUMMARY

This chapter presents a fast-timing electrothermal simulator - ILLIADS-T, which is developed for predicting the temperature distribution and the temperature-sensitive reliability of the VLSI systems.

- The fast-timing simulator used in ILLIADS-T, called ILLIADS, is presented:
  - A generic circuit primitive used in ILLIADS is described.
  - The formation of the state equation and the solutions of the resulting Riccati differential equation are given.
  - The simulation strategy of ILLIADS, including the transistor grouping and merging, is described.
  - The ILLIADS power estimation method for calculating the power of the logic gates is illustrated.
- An incremental strategy for the fast-timing electrothermal simulation is given.
- The details of the tester chip design, calibration and measurement for verifying the accuracy of ILLIADS-T simulation results are provided.
- A number of ILLIADS-T simulation examples are provided. It can be seen that, a logic fault can happen in a circuit and be inadvertently ignored in circuit analysis if the temperature rise or temperature gradient is not considered.

References

[1] B. Chawla, H. Gummel, and P. Kozak, "MOTIS - an MOS timing simulator," *IEEE Transactions on Circuits and Systems*, vol. 22, pp. 901–909, Dec. 1994.

[2] R. Saleh, J. Kleckner, and A. Newton, "Iterated timing analysis and SPLICE1," in *Proceedings of the ACM/IEEE International Conference on Computer-Aided Design*, pp. 139–140, 1983.

[3] J. Ousterhout, "A switch-level timing verifier for digital MOS VLSI," *IEEE Transactions on Computer-Aided Design of Integrated Circuits and Systems*, vol. 4, no. July, pp. 336–349, 1985.

[4] V. B. Rao, *Switch-level timing simulation of MOS VLSI circuits*. PhD thesis, Dept. of Electrical and Computer Engineering, University of Illinois at Urbana-Champaign, 1985.

[5] D. Overhauser, *Fast timing simulation of MOS VLSI circuits*. PhD thesis, Dept. of Electrical and Computer Engineering, University of Illinois at Urbana-Champaign, 1989.

[6] Y. H. Shih, *Computationally efficient methods for accurate timing and reliability simulation of ultra-large MOS circuits*. PhD thesis, Dept. of Electrical and Computer Engineering, University of Illinois at Urbana-Champaign, 1991.

[7] Y. K. Cheng, P. Raha., C. C. Teng, E. Rosenbaum, and S. M. Kang, "ILLIADS-T: An electrothermal timing simulator for temperature-sensitive reliability diagnosis of CMOS VLSI chips," *IEEE Transactions on Computer-Aided Design of Integrated Circuits and Systems*, vol. 17, pp. 668–681, Aug. 1998.

[8] Y. H. Shih, Y. Leblebici, and S. M. Kang, "ILLIADS: A fast timing and reliability simulator for digital MOS circuits," *IEEE Transactions on Computer-Aided Design of Integrated Circuits and Systems*, vol. 12, pp. 1387–1402, Sept. 1993.

[9] A. Dharchoudhury, S. M. Kang, K. H. Kim, and S. H. Lee, "Fast and accurate timing simulation with regionwise quadratic models of MOS I-V characteristics," in *Proceedings of the ACM/IEEE International Conference on Computer-Aided Design*, pp. 208–211, Nov. 1994.

[10] Y. H. Shih and S. M. Kang, "Analytic transient solution of general MOS circuit primitives," *IEEE Transactions on Computer-Aided Design of Integrated Circuits and Systems*, vol. 11, pp. 719–731, June 1992.

[11] W. T. Reid, *Riccati Differential Equations*. New York, NY: Academic Press, 1972.

[12] H. Buchholz, *The Confluent Hypergeometric Function*. New York, NY: Springer Verlag, 1969.

[13] R. E. Tarjan, "Depth first search and linear graph algorithms," *SIAM J. Comput.*, vol. 1, pp. 146–160, June 1972.

[14] J. K. White and A. Sangiovanni-Vincentelli, *Relaxation Techniques for the Simulation of VLSI Circuits*. Norwell, MA: Kluwer, 1987.

[15] S. M. Kang, "Accurate simulation of power dissipation in VLSI circuits," *IEEE Journal of Solid-State Circuits*, vol. 21, pp. 889–891, Oct. 1986.

[16] Y. C. Ju and R. A. Saleh, "Incremental circuit simulation using waveform relaxation," in *Proceedings of the ACM/IEEE Design Automation Conference*, pp. 8–11, 1992.

[17] J. Kim, S. M. Kang, and S. Sapatnekar, "High performance CMOS macromodule layout synthesis," in *Proceedings of the IEEE International Symposium on Circuits and Systems*, pp. 179–182, May 1994.

[18] Microelectronics Center for North Carolina, *VPNR Users Guide*, 1988.

[19] Y. C. Lee, H. T. Ghaffari, and J. M. Segelken, "Internal thermal resistance of a multi-chip packaging design for VLSI based systems," *IEEE Transactions on Components, Hybrids, and Manufacturing Technology*, vol. 12, pp. 163–169, June 1989.

II

THE APPLICATIONS

Chapter 6

TEMPERATURE-DEPENDENT ELECTROMIGRATION RELIABILITY

6.1. MOTIVATION

The device packing density in modern VLSI chips increases steadily. Therefore, the temperature rise in a packaged chip can be dramatic. Because many known IC failure mechanisms are either thermally activated or related, on-chip temperature profile must be predicted prior to any reliability diagnosis and performance analysis. In addition, temperature profile plays an important role in the application of thermal stress evaluation and package design at the chip or the printed-circuit-board (PCB) level. Note that due to large temperature variations across a packaged chip, the assumption of the uniform on-chip temperature usually is not acceptable. In this chapter, one major application of the electrothermal simulation will be presented: temperature-dependent electromigration reliability diagnosis. Other applications such as the cell-based thermal placement and the temperature-driven power/timing analysis will be addressed in subsequent chapters.

In a state-of-the-art chip, the interconnect temperature can rise by as much as 100 °C above the ambient temperature attributed to different heat flow mechanisms. These mechanisms include heat conduction from the substrate, from the nearby interconnects, and heat generated in interconnect itself (*Joule heating*). The temperature affects the diffusion rate of the metal ions because the diffusivity of the ions is exponentially dependent on the temperature. This means that the ions diffuse more rapidly in areas where the temperature is elevated. If there is a temperature gradient in the direction of current flow, an ion flux divergence will be created. Therefore for the locations with higher temperatures, vacancies are likely to form.

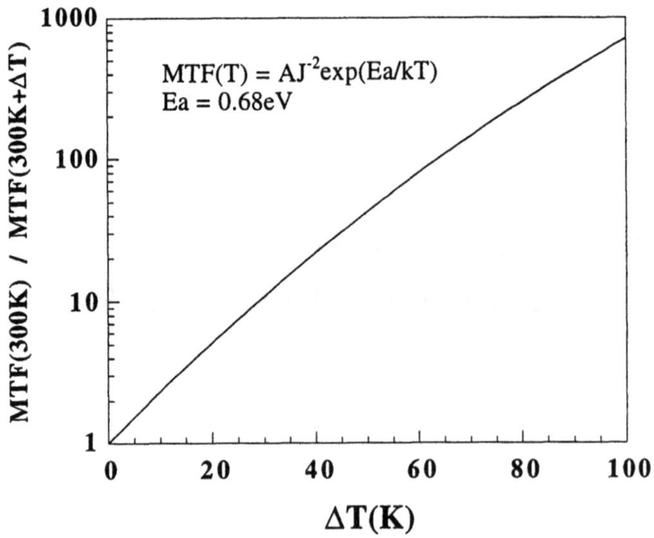

Figure 6.1. The temperature effect on electromigration reliability.

The above temperature-dependent mass transport of metal ions, called *electromigration* (EM), can be described by the well-known Black's equation [1]:

$$\text{MTF} = A \cdot J^{-2} \cdot \exp(E_a/kT) \tag{6.1}$$

In Eq. (6.1), MTF stands for the EM-induced *mean time to failure*, A is a proportionality constant, J is the current density, E_a is the activation energy, k is the Boltzmann's constant, and T is the temperature in degree Kelvin. The physical meaning of the Black's equation Eq. (6.1) will be discussed later. According to Eq. (6.1), the ratio of MTF(T = 300K) over MTF(T = 300K + ΔT) as a function of ΔT is shown in Fig. 6.1.

From Fig. 6.1 it can be seen that if a metal line has a temperature equal to 340 K, its MTF will be twenty times shorter than the MTF when it is subject to the room temperature. It also suggests that neglecting the temperature effect in electromigration failure analysis can substantially overestimate the metal lifetime and lead to unacceptable prediction error.

6.2. ELECTROMIGRATION (EM) PHYSICS

When a metal line is stressed at high current density, mass transport of metal ions driven by the *electron wind* will appear. This mass transport in metal is known as electromigration (EM). In the range of the device operating temperature, since metal ion flux due to the lattice diffusion inside the grain is much smaller than that due to the grain-boundary diffusion, the dominating modes of mass transport are along the grain boundary. The ion flux equation along the grain boundary can be expressed as

$$\vec{J}_{ion} = (Z^* e)\rho n_{gb}(\frac{\omega}{d})(D_{gb}/kT)\vec{J} \tag{6.2}$$

where Z^* is the effective charge of the ion at the grain boundary, ρ is the resistivity, n_{gb} is the ion density in the grain boundary, ω is the average grain boundary width, d is the average grain size, D_{gb} is the ion diffusion coefficient along the grain boundary, and $\vec{J}$ is the current density.

Due to the structural inhomogeneities, temperature gradient, or stress gradient, the ion flux divergence will occur, i.e.,

$$\nabla \cdot \vec{J}_{ion} = -\frac{\partial n}{\partial t} \neq 0. \tag{6.3}$$

The flux divergence can induce damages in the forms of voiding and hillocks. Electromigration-induced voiding can grow and lead to resistance increase or even a catastrophic open of the interconnects. Electromigration-induced hillocks can cause both intralevel and interlevel metal shorts. The times and the locations of void-opens or extrusion-shorts are basically of statistical nature, depending on the spatial distribution of the flux divergence sites. The failure time, of course, is also determined by the magnitude of the flux divergence.

6.2.1 EM LIFETIME DEPENDENCE ON CURRENT DENSITY

The electromigration lifetime dependence on the current density under dc stress was first established by Black [1]. Basically a thermally activated metal ion is acted on by two forces in a metal line: (i) the force created by an electrical field applied to the conductor, and (ii) the rate of momentum exchange between conducting electrons colliding with the activated metal ion. Shielding effects reduce the electrical field effect so that the mass transport of metal ion is mainly driven by the momentum exchange. The rate of mass transport R can be expressed as

$$\begin{aligned} R \propto\ & \text{(electron momentum } P) \\ & \times\text{(number of electrons passing through a unit volume per second } N) \\ & \times\text{(effective target cross section)} \\ & \times\text{(metal activated ion density)} \end{aligned} \tag{6.4}$$

where both the electron momentum P and the number of electrons passing through a unit volume per second N are linearly proportional to the metal current density J. The number of activated ions per unit volume follows the Arrhenius equation as a function of temperature :

$$\text{metal activated ion density} \propto \exp(-E_a/kT). \tag{6.5}$$

Since metal MTF is inversely proportional to the rate of mass transport, Eq. (6.4) can be written as

$$\frac{1}{R} \propto \text{MTF} = A \cdot J^{-2} \cdot \exp(E_a/kT) \tag{6.6}$$

which is the Black's equation Eq. (6.1).

In Black's equation, the inverse-current-square relationship of MTF is suggested. Hinode et al. [2] also found the same result based on both experiments and computer simulation. However, several experimental results have been reported for the values of the current density exponent from 1 to 16, which do not seem to have any agreement on the current density exponent, especially at the high current density range. Sakimoto et al. [3] indicated that the disagreement stems from the lack of accurate temperature data. In the hight current density regime, the self Joule heating increases the metal temperature significantly. If the measurement for the electromigration lifetime is carried out with accurate measurement of the metal temperature, then the current exponent in Black's equation, which is equal to two, is valid at both low and high current density ranges.

6.2.2 EM LIFETIME DEPENDENCE ON CURRENT WAVEFORMS

In the digital circuit environment, the interconnect experiences unidirectional or bidirectional ac current stressing. Many models have been proposed to estimate the EM-induced failure lifetime under such current stress.

UNIDIRECTIONAL AC CURRENT STRESS

Suppose that a metal line is subject to a repetitive pulsed current with a peak value. Intuitively, the MTF of this metal will be longer than the case when it is under the constant dc current stress of this peak value. In the early stage of electromigration research, Towner et al. [4] and Brooke [5] proposed an empirical electromigration failure model for the unidirectional pulsed current stress using experimental results:

$$\text{MTF}_{\text{pulse}} = A \cdot J_{avg}^{-2} \cdot \exp(E_a/kT) \tag{6.7}$$

where $\text{MTF}_{\text{pulse}}$ is the EM-induced MTF under unidirectional pulsed current stress, and J_{avg} is the average current density defined as

$$J_{avg} = \frac{1}{t_p} \int_0^{t_p} J(t)\, dt. \tag{6.8}$$

Equation (6.7) is usually referred to as the *average current model.*

Later, Maiz [6] applied the kinetics of material accumulation/depletion to approximate the electromigration failure model under pulsed current stress. In addition, other models have been proposed such as the *vacancy supersaturation model* by Clement [7] and the *defect relaxation model* by Tao et al.[8]. Although different theoretical models were proposed, they all arrived at the same conclusion and validated the average current model for unidirectional pulsed current stress. Here, the defect relaxation model [8] will be used to explain the average current model. Assume that δ represents the volume of voiding in the interconnects. The increase of δ in the unit time is proportional to the product of vacancy concentration n and current density J. Hence,

$$\frac{d\delta(t)}{dt} = R(\delta) \cdot n \cdot J. \tag{6.9}$$

The proportionality constant R is a function of δ. Suppose that at time MTF, the $\delta(t)$ reaches some critical value δ_c and causes the failure of the interconnect. The above relation can be written as

$$\int_0^{\mathrm{MTF}} n \cdot J dt = \int_0^{\delta_c} \frac{d\delta(t)}{R(\delta)} \equiv K \tag{6.10}$$

where K is a constant. Assuming that the vacancy relaxation time is τ and the rate of vacancy generation is proportional to $|J|$, then

$$\frac{dn}{dt} = -\frac{n}{\tau} + \alpha|J|. \tag{6.11}$$

If J is a dc current, by solving Eq. (6.11), the vacancy concentration can be obtained as follows

$$n = \tau\alpha J. \tag{6.12}$$

Substituting Eq. (6.12) into Eq. (6.10), then the $\mathrm{MTF_{dc}}$ is

$$\mathrm{MTF_{dc}} = \frac{K}{\tau\alpha J^2}. \tag{6.13}$$

If we set

$$\frac{K}{\tau\alpha} = A \cdot \exp(E_a/kT) \tag{6.14}$$

the MTF under dc stress can have a simple form:

$$\mathrm{MTF_{dc}} = A \cdot J^{-2} \cdot \exp(E_a/kT). \tag{6.15}$$

The above equation exactly matches the Black's equation. Now considering a unidirectional pulsed current where the peak value is J and the duty factor is

D, solve Eq. (6.11) to obtain the vacancy concentration and apply the solution to Eq. (6.10) again. The $\mathrm{MTF}_{\mathrm{pulse}}$ can be obtained as

$$\mathrm{MTF}_{\mathrm{pulse}} = \frac{\mathrm{MTF}_{\mathrm{dc}}}{D\left[1 - \frac{(1-e^{-aD})(1-e^{-a(1-D)})}{aD(1-e^{-a})}\right]} \tag{6.16}$$

where $a = 1/(f\tau)$, and f is the frequency of the dc pulse. The vacancy relaxation time τ is approximately equal to 1 ms. When the operating frequency is much higher than τ, i.e., $f \gg 1$ KHz, then Eq. (6.16) can be approximated as

$$\mathrm{MTF}_{\mathrm{pulse}} = \frac{\mathrm{MTF}_{\mathrm{dc}}}{D^2} = A \cdot J_{avg}^{-2} \cdot \exp(E_a/kT). \tag{6.17}$$

For an arbitrary unidirectional ac current waveform in the high frequency region, Eq. (6.7) is still valid [8, 9].

BIDIRECTIONAL AC CURRENT STRESS

The signal lines between logic gates carry bidirectional current in the CMOS circuits. It is thus essential to assess the reliability of a metalization system under bidirectional current stress. It has been observed that the electromigration damage incurred by the positive (forward) current stress can be partially healed by the following negative (reverse) current stress [8, 10, 11, 12, 13]. Ting et al. [12] have empirically shown that

$$\mathrm{MTF}_{\mathrm{bidi}} = A \cdot J_{eff}^{-2} \cdot \exp(E_a/kT). \tag{6.18}$$

This model is called the *average current recovery* model. The effective current density J_{eff} is defined as

$$J_{eff} = |\bar{J}^{+}| - \gamma|\bar{J}^{-}| \tag{6.19}$$

where $\bar{J}^{+}$ and $\bar{J}^{-}$ are the average current density including only the positive and the negative current, respectively. In Eq. (6.19), $|\bar{J}^{+}| \geq |\bar{J}^{-}|$ is assumed. Considering the bidirectional pulse waveform shown in Fig. 6.2, for instance, we have

$$|\bar{J}^{+}| = \frac{1}{\tau_{total}} \sum_i J_{2i-1}\tau_{2i-1} \tag{6.20}$$

$$|\bar{J}^{-}| = \frac{1}{\tau_{total}} \sum_i J_{2i}\tau_{2i}. \tag{6.21}$$

In Eq. (6.19), γ represents the degree of damage recovery due to the current with opposite polarity. It has to be in the range [0,1]. In the extreme cases, if $\gamma = 0$, there is no healing effect in ac stressing, whereas if $\gamma = 1$, there is perfect healing during the negative pulse. Usually the value of γ is around 0.9. The validity of the average current model has been shown in [12].

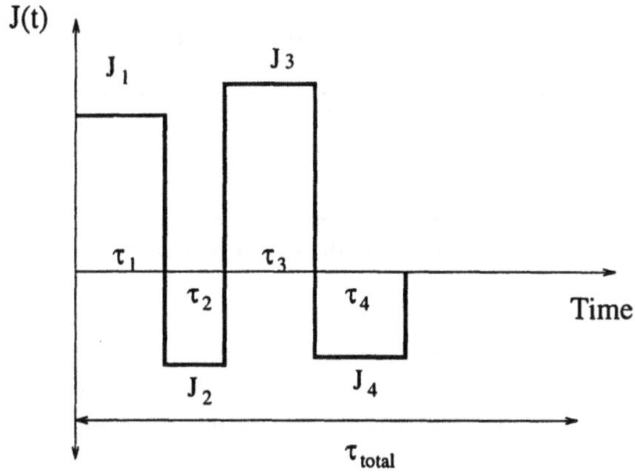

Figure 6.2. An example of a bidirectional pulsed current density waveform.

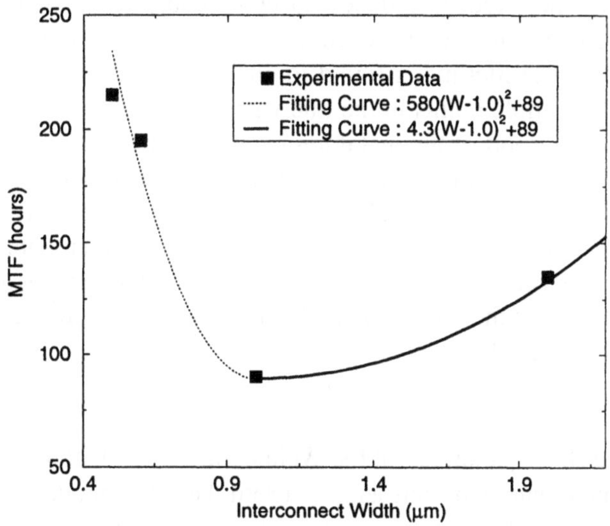

Figure 6.3. Electromigration MTF as a function of interconnect width [14].

6.2.3 EM LIFETIME DEPENDENCE ON INTERCONNECT WIDTH AND LENGTH

The electromigration lifetime as a function of the interconnect width is shown in Fig. 6.3. For interconnects with the width larger than the metal grain size (Figure 6.4(a)), the major mechanism of electromigration is the grain-boundary diffusion occurring in the *triple points*. Since the EM-induced open-circuit failure in the metal line is along its width, an increase of lifetime

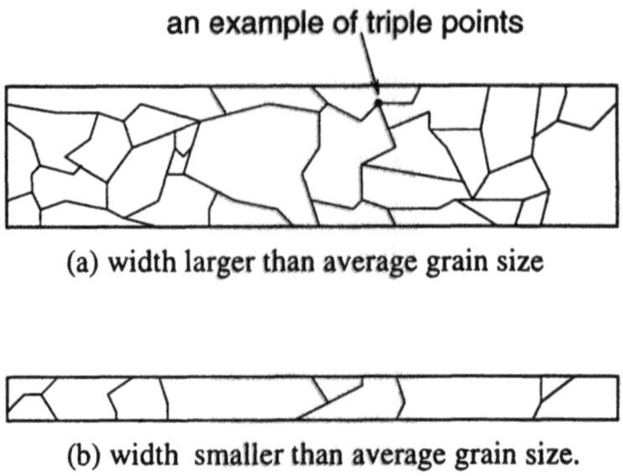

Figure 6.4. Microstructure of the interconnects.

with increasing line width is expected because the probability of making a wider interconnect open is smaller.

However, as the width is reduced to comparable to or smaller than the mean grain size, the metal line becomes more or less a *bamboo structure* (Figure 6.4(b)). The lifetime is found to either level off or increase because the number of the triple points in the line with bamboo structure is much less. Thus, the mass transport of the metal ion is more difficult to occur.

The electromigration lifetime dependence on the metal line length has been investigated in many papers [14, 15, 16, 17]. The simplest analysis method is to use the *series model* [17, 18]. The series model treats an interconnect of length L as a series connection of L unit lines. Assuming that the electromigration failure model follows the *Weibull distribution,* the MTF of an interconnect of length L can be derived as

$$\mathrm{MTF}_L = L^{-1/m} \cdot \mathrm{MTF}_{unit} \tag{6.22}$$

where m is the shape parameter of the Weibull distribution, and MTF_{unit} is the MTF of the interconnect of unit length. From the series model it can be seen that the MTF approaches zero as the metal line is long enough.

The typical results of the electromigration lifetime dependence on the interconnect length are shown in Fig. 6.5 [15, 16]. The MTF decreases rapidly with increasing line length and then reaches a saturation value beyond a critical length. It clearly contradicts the series model prediction. Note that the series model treats every unit of interconnects as independent random variables, which is not true in the physical mechanism behind electromigration.

In thin-film interconnects, the EM-induced damage due to flux divergence can occur at any intrinsic defects (e.g., dislocations of grain boundary) or extrinsic defects (e.g., process-induced defects [16]). In essence, the EM-induced failure lifetime is dependent on the *most severe* defect in the whole

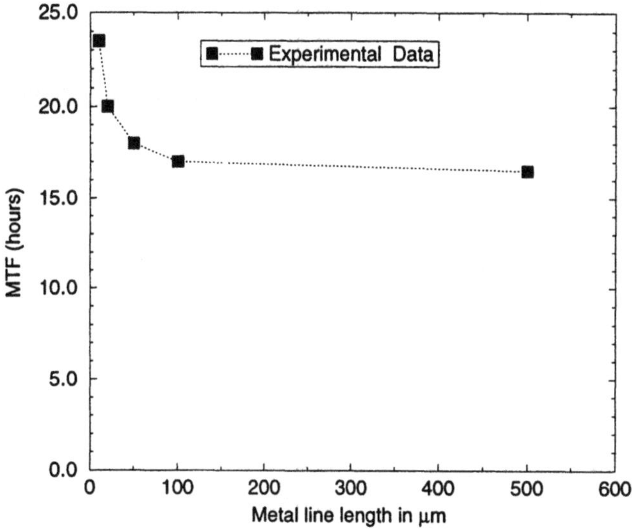

Figure 6.5. Electromigration MTF as a function of interconnect length.

interconnect, rather than on the number of the defects with the severity beyond a certain level. The possibility of finding more severe defects increases with the interconnect length, so that the lifetime of longer interconnects is shorter. However, because of the stability of processing and the characteristics of the material structure, the severity of extrinsic and intrinsic defects has an upper bound. Therefore, the EM-induced lifetime levels off when the interconnect length is greater than the critical length.

6.2.4 EM MODEL USED IN THE BOOK

There is still no consensus on the theory of the electromigration phenomenon. The analytical failure model for electromigration used in this book mainly adopts the average current recovery model in Eq. (6.18), which is derived from the theoretical and experimental works and is believed to be correct.

A more general form of the average current recovery model describing the MTF of an interconnect can be expressed as:

$$\mathrm{MTF}_{l,w} = A(l,w) \cdot J_{eff}{}^{-n} \cdot \exp(E_a/kT) \qquad (6.23)$$

where $\mathrm{MTF}_{l,w}$ is the MTF of the interconnect with length l and width w. The proportionality constant A is now a function of l and w.

6.3. EM SIMULATION: AN OVERVIEW

In a VLSI system, the development of computer programs that are capable of simulating a large number of interconnects and accurately predicting the

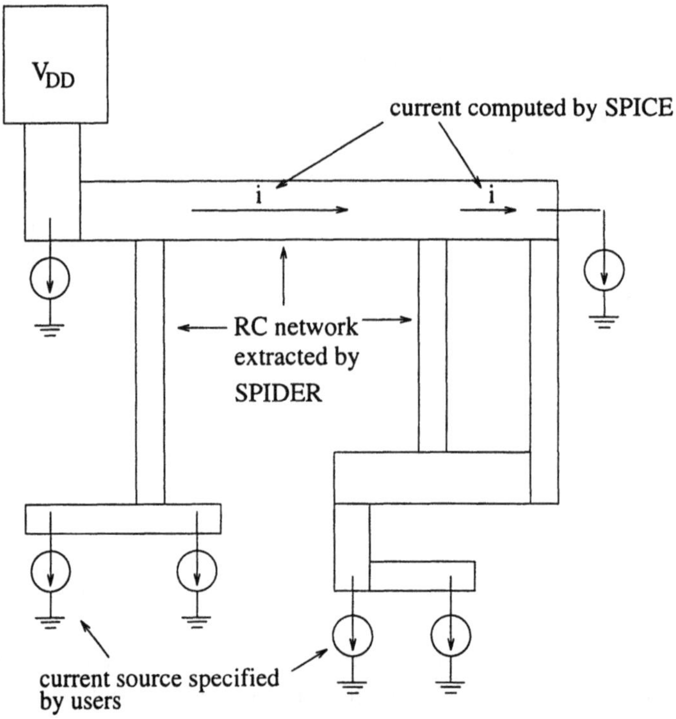

Figure 6.6. SPIDER [19] for the simulation of interconnect reliability.

electromigration lifetime has been a constant challenge. SPIDER [19] was one of the first tools developed for the simulation of interconnect reliability. In SPIDER, users select a portion of the metal bus and specify the current sources to load the metal bus at specified contact points, as shown in Fig. 6.6. Then SPIDER extracts an equivalent RC network for the metal bus and simulates the network using SPICE [20]. The current waveforms of every segment of interconnects can be obtained and used for the design and analysis of the metal system.

Another electromigration reliability CAD tool, called RELIANT [21], was developed in 1989. RELIANT extracts the interconnect network and the transistors from the circuit layout. A switch-level simulator, CURRANT [22], is then employed to calculate the current waveform of each interconnect. In CURRANT, each transistor is substituted by a resistor that represents the source-to-drain resistance. An example demonstrating the substitution for a 2-input logic NAND gate is shown in Fig. 6.7. The value of the resistance depends on whether the transistor is on or off. Interconnects are represented by resistors and capacitors. Next, CURRANT analyzes the mapped RC network to determine the delay and the transient current waveform of each interconnect.

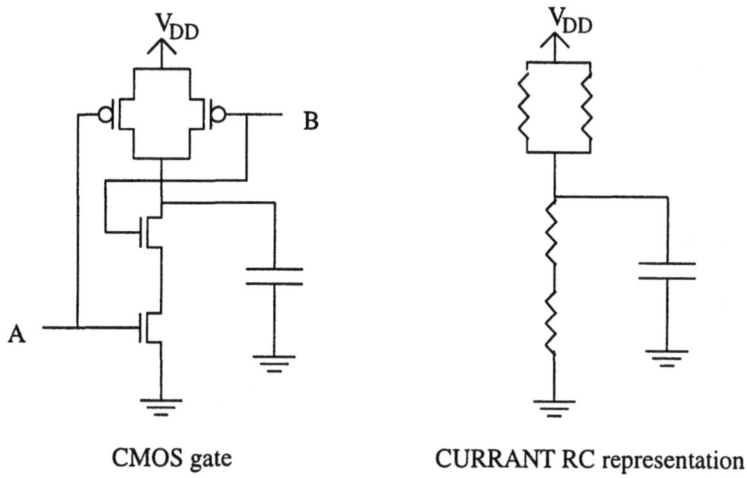

Figure 6.7. CURRANT representation of a 2-input NAND gate.

RELIANT can handle large circuits, but the inaccurate switch-level simulator may cause significant errors, especially for the deep submicron technology.

The Berkeley reliability tool (BERT) [23] is an integrated reliability tool. It can simulate the circuit reliability of the hot-carrier effect, the time-dependent dielectric breakdown, and the electromigration. There are two modes of operation in BERT electromigration simulation. In the first mode, users provide a SPICE input file without any layout information. Then BERT generates a layout advisory based on the current flowing in the circuit and the user-specified reliability requirements. This advisory can be used for the design of an electromigration reliable interconnect system. The second mode is to simulate the electromigration failure rate for the given circuit layout. In this mode, BERT first extracts the circuit netlist from the target layout and generates a SPICE input deck. At the same time, it produces a geometry description file for all interconnects, vias, and metal-to-diffusion contacts. Next, BERT runs SPICE circuit simulation to obtain the current waveform of the transistors. The postprocessor of BERT calculates the failure rate as a function of the data in the geometry description file and the SPICE simulation result. Compared with RELIANT, BERT is much more accurate. However, it is computationally expensive due to the amount of CPU time that SPICE requires. Two other reliability simulators, RELIC [24] and RELY [25], employ the approach similar to BERT for the electromigration simulation.

All of the above electromigration CAD tools require users to provide either the input current load of the interconnects or the input vectors of the circuit. Their simulation results are strongly dependent on the specified input patterns. An alternative approach is to use the *probabilistic* simulation [26, 27, 28] to

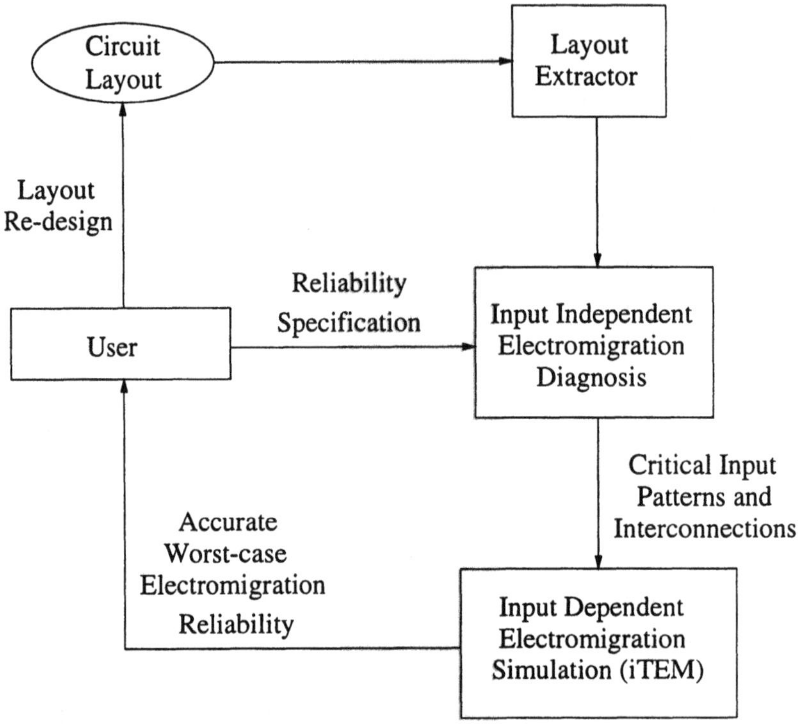

Figure 6.8. A hierarchical environment for interconnect EM reliability diagnosis.

get the *average* current stress in interconnections without knowing the input patterns. However, the signal correlations between internal nodes in the probabilistic simulation need to be properly handled for accuracy. Since handling the signal correlation is difficult and expensive, the accuracy is usually traded for speed. Moreover, the average-case analysis is not able to predict the worst-case scenario, which is important in the design-for-reliability paradigm.

A hierarchical electromigration reliability diagnosis tool [29] was developed to resolve the speed and accuracy problems in above approaches. Two levels of diagnosis hierarchy are used as shown in Fig. 6.8. The top hierarchy is the input pattern-independent electromigration diagnosis. It can quickly identify the critical interconnects with potential electromigration reliability problems. The corresponding input patterns which cause the worst-case current stress to each critical interconnect are also found. The problem size of the worst-case electromigration reliability analysis is thus significantly reduced and becomes tractable. Designers can focus on those critical interconnects and feed the corresponding critical input patterns into the second hierarchy: the input pattern-dependent electromigration reliability simulation. The simulation tool is called iTEM [30].

The acronym iTEM stands for the **I**llinois **T**emperature-dependent **E**lectro-**M**igration diagnosis tool. It accurately computes the EM-induced failure of the interconnect system. One important feature of iTEM is that, in addition to the current density and the interconnect geometry, it takes into account the steady-state temperature of every interconnect under the given operating conditions. As mentioned earlier, the temperature effect is critical to the accuracy of the electromigration MTF estimation. In the following of the chapter, the temperature-dependent electromigration reliability diagnosis flow in iTEM will be examined in detail.

6.4. ITEM: A TEMPERATURE-DEPENDENT EM DIAGNOSIS TOOL

In order to estimate the interconnect temperature, the first step is to find the substrate temperature distribution based on which the individual interconnect temperature can be calculated subsequently. Different approaches for finding the substrate temperature were presented and discussed in Chapter 4. Among all approaches, the fast timing electrothermal simulator ILLIADS-T [31] was mainly designed for the VLSI systems. Since iTEM also aims to predict the electromigration problems at the VLSI level, ILLIADS-T is employed as its simulation engine for substrate temperature calculation. The iTEM simulation flow is shown in Fig. 6.9.

After the ILLIADS-T electrothermal simulation, iTEM extracts the power and ground buses from the layout and identifies the transistors in contact with the ground/power buses. The correspondence between the bus and the transistors that are connected to it is also identified concurrently. The extraction procedures are similar to those used in [32]. Next, iTEM extracts the resistive networks from the buses and builds the admittance matrices for the networks. The currents drawn from the transistors connected to the buses serve as the current sources of the networks, and the admittance matrices of the network are solved. At this stage, the current waveform of each metal rectangle, via and metal-diffusion contact is found for the ground and power buses.

6.4.1 INTERCONNECT TEMPERATURE ESTIMATION

The previous discussion described the procedures for finding the substrate temperature profile and the interconnect current waveform for the ground and power buses. The next step is to estimate the interconnect temperature in order to accurately predict the EM-induced MTF. The following assumptions for interconnect temperature estimation are made in iTEM:

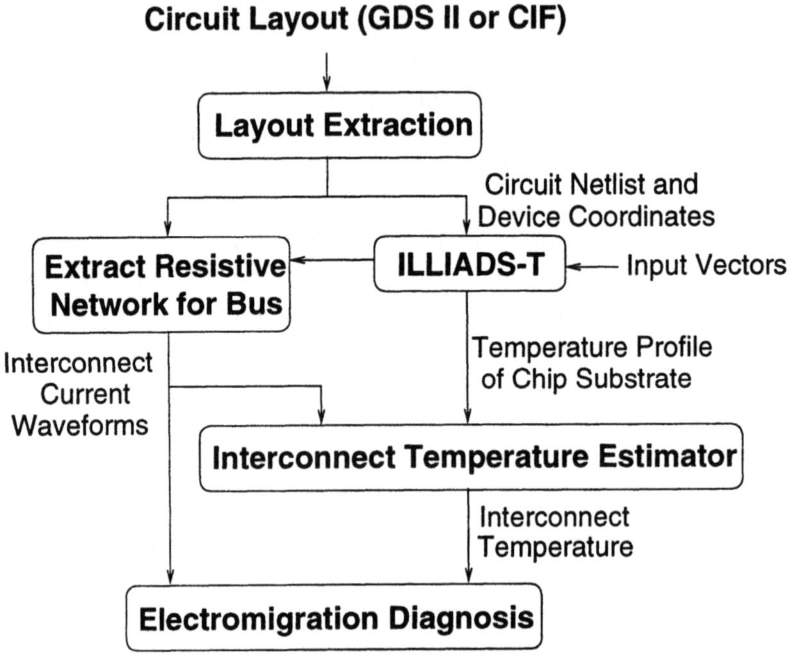

Figure 6.9. Simulation flowchart of iTEM.

1. Three phenomena may raise the interconnect temperature: self Joule heating, heat conduction from the substrate, and heat conduction from nearby interconnects. The heat conduction from nearby interconnects is assumed insignificant and ignored in the analysis.

2. It is assumed that all the heat generated in the interconnect is conducted away by the silicon substrate.

3. The thermal conductivity dependence on temperature is assumed weak and is ignored.

The first assumption is generally true because typically the heat contributed from the nearby interconnects is mush less than from the substrate and Joule heating. The second assumption is justified because for a packaged chip, the top of the interconnects is covered with a thermal insulation layer, such as overglass, silicon dioxide, or silicon nitride. This top layer is thicker than the silicon dioxide layer separating the substrate and interconnects. As a result, almost all of the interconnect heat flow travels through the silicon substrate which has a high thermal conductivity. The third assumption is valid if the on-chip temperature gradient is not too big. If it is not the case, the temperature dependence of the thermal conductivity needs to be considered in the analysis.

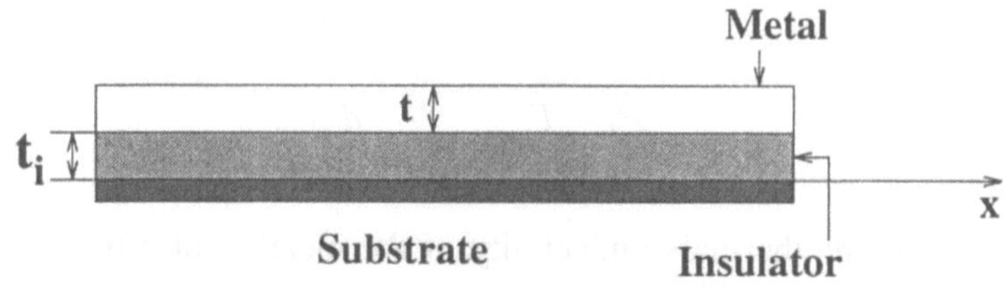

Figure 6.10. The interconnect on insulator structure.

6.4.2 ANALYTICAL MODEL OF INTERCONNECT THERMAL SYSTEM

To estimate the interconnect temperature, the following case shown in Fig. 6.10 is first examined. This structure consists of a long metal wire and an insulator on the substrate. The width and thickness of the metal line is w and t respectively, and the thickness of the insulator is t_i. The resistivity ρ of the metal line is a function of temperature T as follows:

$$\rho = \rho_0 \cdot [1 + \beta \cdot T] \tag{6.24}$$

where ρ_0 is the resistivity at 0 °C and β is the temperature coefficient of resistivity. The Joule heating P_j generated by a segment of interconnect with length Δx is

$$P_j = I^2 \Delta R = I^2 \rho \frac{\Delta x}{wt}. \tag{6.25}$$

The heat P_i conducted to the substrate is

$$P_i = \frac{T_m - T_s}{R_{th}} = \frac{K_{i,eff} w \Delta x}{t_i} \hat{T_m} \tag{6.26}$$

where R_{th} is the thermal resistance between the interconnect and the silicon, T_s is the substrate temperature, T_m is the metal temperature, and $\hat{T_m} = T_m - T_s$.

In Eq. (6.26), $K_{i,eff}$ is the effective thermal conductivity of the insulator taking into consideration the deviation from one-dimensional heat flow. Given that the thermal conductivity in the bulk of the insulator is K_i, the ratio of $(K_{i,eff}/K_i)$ is larger than 1 and increases when (t_i/w) decreases, because the heat fringing effect increases. By using the Schwarz-Christoffel conformal transformation [33, 34], the ratio of $(K_{i,eff}/K_i)$ can be approximated by

$$\frac{K_{i,eff}}{K_i} = \left(1 + 0.88 \frac{t_i}{w}\right) \tag{6.27}$$

within 3% of error if $w > 0.4t_i$. The heat diffusion equation along the x-axis in Fig. 6.10 is

$$Q = \frac{P_j - P_i}{wt\Delta x} = -K_m \frac{d^2\hat{T_m}}{dx^2} \tag{6.28}$$

where K_m is the thermal conductivity of the metal. Substituting Eqs. (6.25) and (6.26) into Eq. (6.28) results in

$$K_m \frac{d^2\hat{T_m}}{dx^2} + \frac{I^2\rho}{w^2t^2} - \frac{K_{i,eff}}{t \cdot t_i}\hat{T_m} = 0. \tag{6.29}$$

The resistivity of the interconnect is temperature dependent as described in Eq. (6.24). When the temperature dependence of the resistivity is included, we have

$$\frac{d^2\hat{T_m}}{dx^2} - \frac{1}{K_m}\left(\frac{K_{i,eff}}{t \cdot t_i} - \frac{I^2\rho_0\beta}{w^2t^2}\right)\hat{T_m} + \frac{I^2\rho_0(1+\beta\hat{T_m})}{K_m w^2t^2} = 0. \tag{6.30}$$

If the metal length is long enough, $\frac{d^2\hat{T_m}}{dx^2} = 0$ at steady state. Hence, the metal temperature T_m can be presented as [33, 35]:

$$T_m = T_s + \frac{J^2\rho_0(1+\beta T_s)}{\frac{K_{i,eff}}{t \cdot t_i} - J^2\rho_0\beta} \tag{6.31}$$

where $J = I/(w \cdot t)$ is the current density.

A 3-D numerical thermal simulator was used to verify the above equation. Two cases were simulated, In the first case, $K_{i,eff}$ in Eq. (6.31) was replaced by K_i (i.e., the 1-D model). In the second case, Eq. (6.27) was used for $K_{i,eff}$ in Eq. (6.31). The results are shown in Fig. 6.11, where it can be seen that (i) the error resulting from the 1-D model is not acceptable, and (ii) Eq. (6.31) matches quite well with the 3-D simulation if Eq. (6.27) is used for $K_{i,eff}$.

6.4.3 LUMPED MODEL OF INTERCONNECT THERMAL SYSTEM

A 3-D interconnect structure (Fig. 6.12(a)) carrying 9 MA/cm^2 current was simulated by a 3-D numerical thermal simulator with the substrate maintained at 350 K and the bond pads at 300 K. The result is given in Fig. 6.12(b). It shows that for the locations in the interconnect that are at least one thermal diffusion length away from the metal-diffusion contacts or pads, their temperatures are well-modeled by Eq. (6.31). However, the interconnect temperature of the points close to the contacts and pads does not follow Eq. (6.31) since the contacts and pads are good heat sinks.

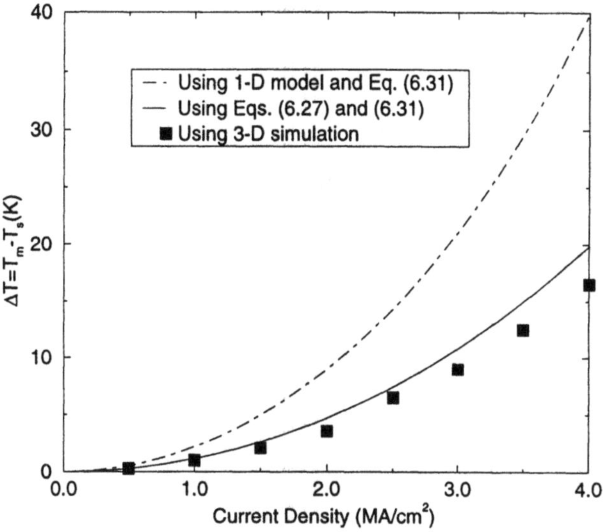

Figure 6.11. ΔT is a function of metal current density. The following parameters values are used to generate the data: $t_i = 2\ \mu$m, $t = 0.5\ \mu$m, $w = 2\ \mu$m, $\rho_0 = 3.6 \times 10^{-6}\ \Omega\cdot$ cm, $\beta = 4.04 \times 10^{-3}$ K^{-1}, $K_i = 1.835$ W/(K·m), and T_s=300 K.

From the above observation, it is clear that Eq. (6.31) can overestimate the interconnect temperature in many situations. On the other hand, the boundary conditions of the full-chip interconnect thermal system are so complicated that it is impossible to solve the heat diffusion equation analytically like the methods used in [33, 35]. Instead, a *lumped model* to estimate the interconnect temperature was proposed in iTEM.

Consider the structure of a metal interconnect on an insulator as shown in Fig. 6.13. The width and thickness of the metal line is w and t respectively, the insulator thickness is t_i, and the current flowing through the metal is I. For a segment of interconnect with length Δx, the local thermal system can be mapped into the equivalent electrical network shown in Fig. 6.13 by using the thermal-electrical analogy described in Chapter 4. In Fig. 6.13, V_s and V_m are the substrate and metal temperatures respectively. The Joule heating of the metal comprises two elements: I_R and $I_{\Delta R}$. The first element is the *constant current source* in Fig. 6.13, which is the primary contributor of Joule heating and is calculated as

$$I_R = I^2 \rho_0 \frac{\Delta x}{w \cdot t}. \tag{6.32}$$

As mentioned before, the metal resistivity increases with temperature (Eq. (6.24)). Therefore the second element of Joule heating can be represented by a *voltage-*

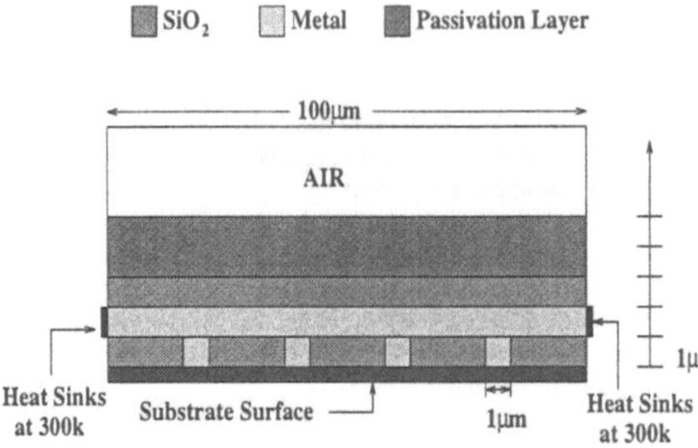

(a) Cross-sectional view

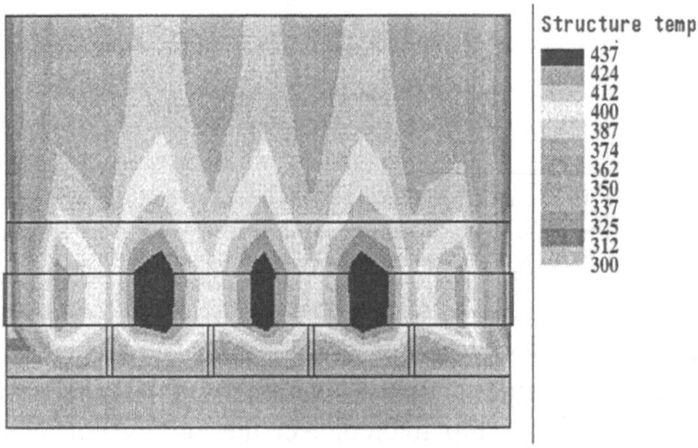

(b) Temperature distribution

Figure 6.12. (a) Interconnect with contacts to substrate, and (b) the corresponding temperature distribution from 3-D thermal simulation. Note that the interconnect temperature is reduced near the contacts and bond pads.

dependent current source that is due to the resistivity increase caused by the interconnect temperature rise:

$$I_{\Delta R} = \left(I^2 \rho_0 \beta \frac{\Delta x}{w \cdot t} \right) \cdot V_m. \tag{6.33}$$

In Fig. 6.13, R_m is the thermal resistor of the metal given by

$$R_m = \frac{0.5 \Delta x}{K_m \cdot w \cdot t} \tag{6.34}$$

where K_m is the thermal conductivity of the metal. The thermal conduction path between the interconnect and the chip substrate is described by R_i. If

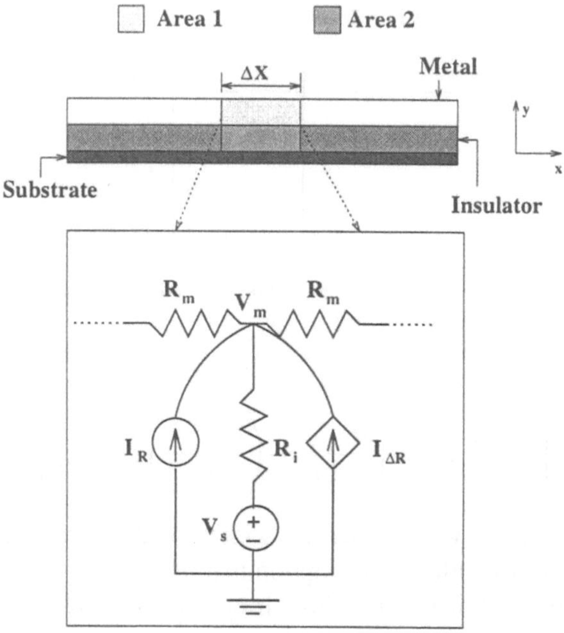

Figure 6.13. A lumped model of the interconnect thermal system.

the material between the metal and the substrate (i.e., Area 2 in Fig. 6.13) is insulator, then

$$R_i = \frac{t_i}{K_{i,eff} \cdot \Delta x \cdot w}. \tag{6.35}$$

If the material is metal, i.e, contacts, then

$$R_i = \frac{t_i}{K_m \cdot \Delta x \cdot w}. \tag{6.36}$$

Based on the lumped model, let us again examine the long metal line structure shown in Fig. 6.10. Since $R_m \approx \infty$ ($\Delta x \approx \infty$), the interconnect temperature V_m is

$$V_m = V_s + (I_R + I_{\Delta R}) \cdot R_i. \tag{6.37}$$

After substituting Eqs. (6.32), (6.33), and (6.35) into the above equation, an expression for V_m (i.e., T_m) will be obtained, which is the same as Eq. (6.31).

An important shape of the interconnect that needs to be taken into account is the right-angle bend (L-shape), as shown in Fig. 6.14. A two-dimensional analytical formula is used to approximate the thermal resistance of the corner

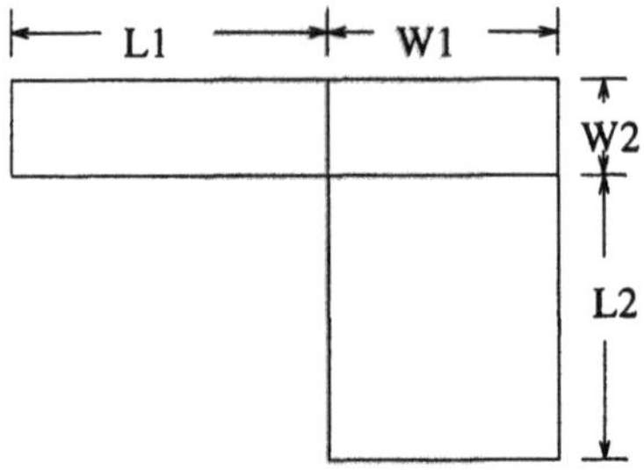

Figure 6.14. A right-angle bend conductor.

rectangle [37]:

$$R_m = \frac{0.5}{K_m \cdot t}\left[\frac{1}{a} - \frac{2}{\pi}\ln\left(\frac{4a}{a^2+1}\right) + \left(\frac{a^2-1}{\pi a}\right)\cos^{-1}\left(\frac{a^2-1}{a^2+1}\right)\right] \quad (6.38)$$

where a is the ratio of wide-to-narrow widths of the corner rectangle. (In Fig. 6.14, $a = w_1/w_2$ with the assumption that $w_1 \geq w_2$.) A similar equation can be derived for the T-shape. For those irregularly shaped conductors requiring the high accuracy of the heat resistance calculation, the finite-difference or finite-element method can be used [37]. For the heat interaction between the interconnects in different layers, iTEM only considers the heat path through the vias. An example of the lumped model for the interconnect thermal system near a via is shown in Fig. 6.15.

The above lumped thermal models were verified by the accurate 3-D numerical thermal simulation. The first structure tested is a long interconnect with four contacts to the substrate as shown in Fig. 6.16(a). The simulation parameters are the same as in Fig. 6.11. The substrate temperature is maintained at 300 K. Figure 6.16(b) shows the simulated interconnect temperature distribution along the x-axis when the interconnect current densities are 2 MA/cm^2 and 3 MA/cm^2. The second structure is a two-layer metal structure. Metal 1 has two contacts to the substrate, and Metal 2 has one via to Metal 1 as shown in Fig. 6.17(a). The thickness between different layers is 1 μm. The substrate temperature is maintained at 300 K and the current density at both metal layers is 3 MA/cm^2. The simulation results are shown Fig. 6.17(b). In both examples, the temperature difference between the simulation results using the lumped model and the 3-D simulation is at most 1 K. This difference is mostly due to the inherent error of $K_{i,eff}$. (See Fig. 6.11.)

Using the lumped thermal model, the procedure of interconnect temperature estimation in iTEM is shown in Fig. 6.18. The first step is to partition the interconnect layout according to the geometry . An example is given in Fig. 6.19. The partitioning rule is almost the same as that used in the parasitic resistance extraction [38]. After partitioning, every segment of the intercon-

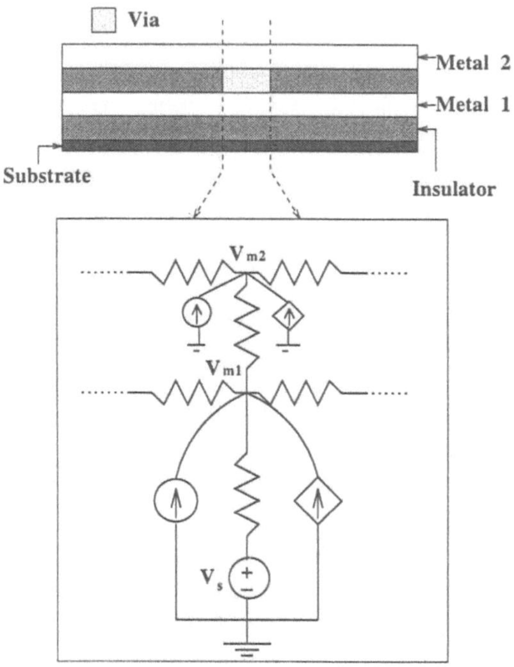

Figure 6.15. A lumped model of the interconnect thermal system near a via.

nects is mapped into a thermal resistor as shown in Fig. 6.13. Thus, a thermal resistive network describing the interconnect thermal system is obtained. Next, the admittance matrix of the thermal network is formed and the interconnect temperatures are solved.

Note that the contacts in one source/drain area are usually very close to each other. To reduce the number of nodes in the interconnect thermal resistive network without loss of accuracy, the contacts that belong to one diffusion region can be heuristically grouped into one segment. For instance, the entire segment ΔX with four contacts in Fig. 6.20 will be mapped into one lumped thermal resistive network since the four contacts locate in the same diffusion area. The materials below the interconnect with multiple contacts are not homogeneous. The *equivalent thermal resistance* method [39] can be applied to describe the non-homogeneous heat path between the metal and the substrate. For example, in Fig. 6.20:

$$R_{eq} = \frac{t_i}{K_{eq} \cdot \Delta x \cdot w} \tag{6.39}$$

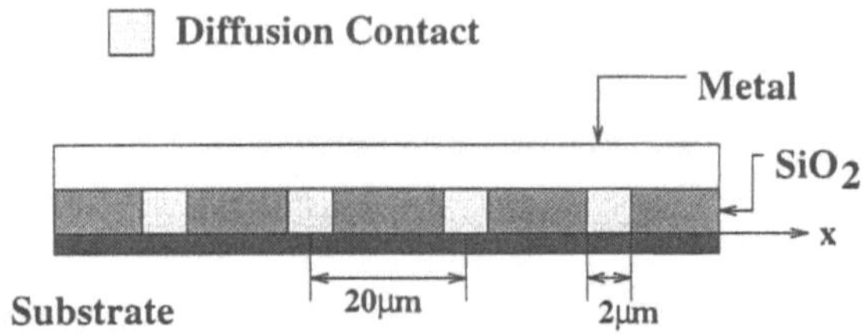

(a) A simulated structure

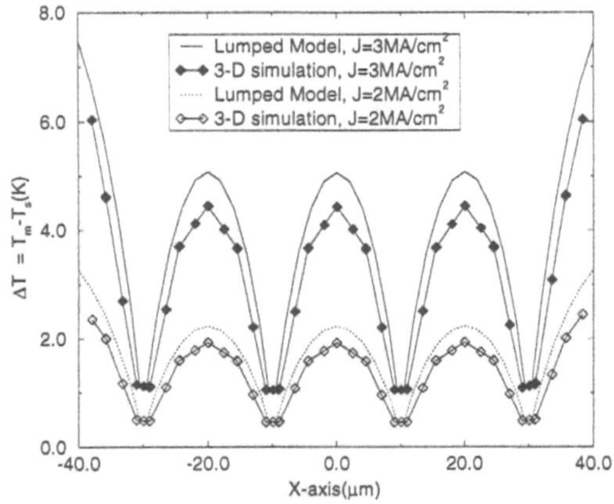

(b) Simulation results

Figure 6.16. (a) Simulated interconnect structure with four contacts to substrate. (b) Comparison of the thermal simulation results using lumped thermal model and 3-D simulation.

where

$$K_{eq} = \frac{K_m(\text{area of contacts}) + K_{i,eff}(\Delta x \cdot w - \text{area of contacts})}{\Delta x \cdot w}. \quad (6.40)$$

The same structure shown in Fig. 6.16(a) is again simulated, but the distance between contacts is reduced to 2 μm. The current density is assumed 3 MA/cm^2. Two partitioning strategies are compared. One is to partition the four contacts into different segments as in Fig. 6.21(a). The other is to lump the contacts together and use the concept of equivalent thermal resistance as in Fig. 6.21(b). The result in Fig. 6.22 indicates that lumping the close contacts together is a proper approximation, which substantially reduces the complexity of the thermal resistive network.

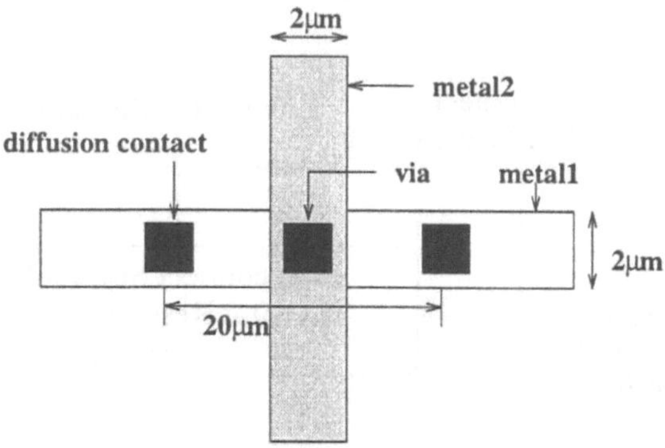

(a) A simulated structure

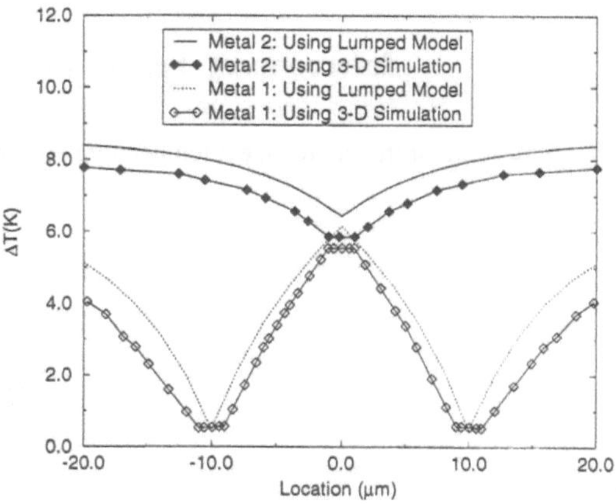

(b) Simulation results

Figure 6.17. (a) Simulated multi-layered interconnect structure. (b) Comparison of thermal simulation results using lumped thermal model and 3-D simulation.

6.4.4 ITEM SIMULATION EXAMPLES

Figure 6.23 is the layout of the 10-bit negative adder used as a test circuit with the input signal frequency of 300 MHz. Its power/ground bus layout is shown in Fig. 6.24. Based on the iTEM simulation, the electromigration diagnosis result is shown in Fig. 6.25 for the region within the box in Fig. 6.23. The number marked in each metal rectangle and contact is the predicted MTF in hours. Several metal lines have an "Inf" MTF since there exist transistors not switching during the simulation period.

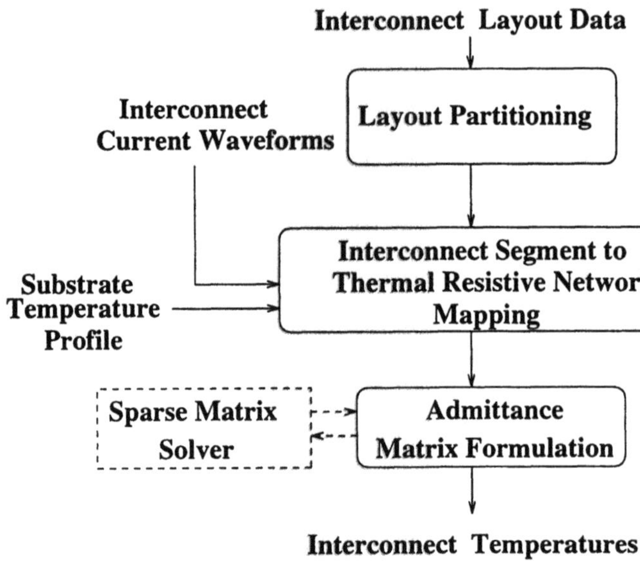

Figure 6.18. Procedure of the interconnect temperature estimator.

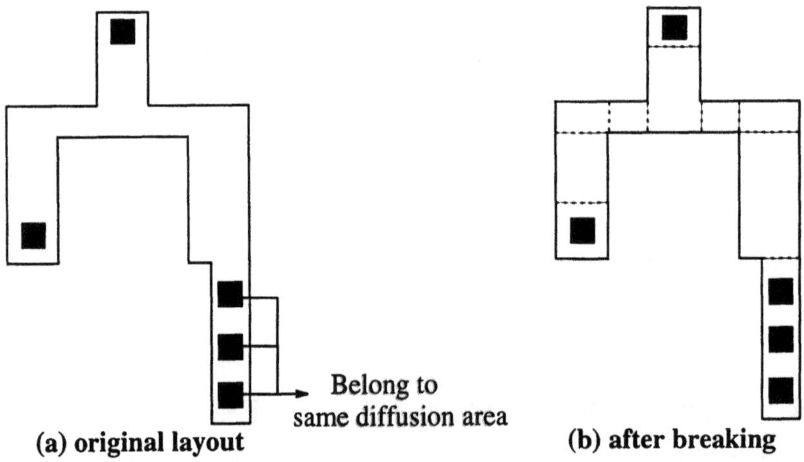

Figure 6.19. Example of partitioning the interconnect layout in the interconnect temperature estimator.

Next, a large circuit containing about 110k transistors is simulated. It is an 8×8 2-D discrete cosine transformation (DCT) chip [40] and its power bus

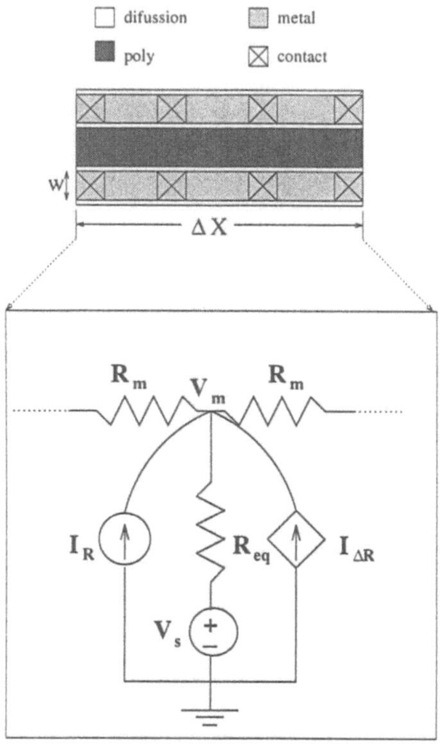

Figure 6.20. The lumped thermal model for a transistor with multiple contacts.

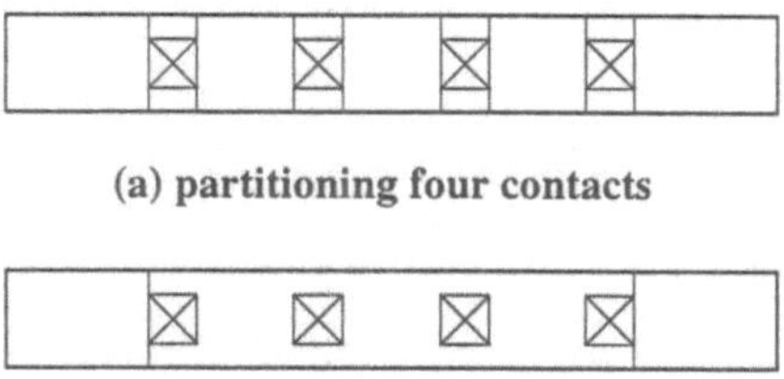

(a) partitioning four contacts

(b) lumping four contacts together

Figure 6.21. Strategies for grouping contacts that are close to each other.

layout is shown in Fig. 6.26. The electromigration diagnosis result for a small region of the layout is shown in Fig. 6.27.

Finally, the detailed iTEM simulation results for four different circuits are shown in Table 6.1. Twenty input vectors are input into each testing circuit. The MTF shown in the table is the shortest MTF among all interconnects in the circuit. Note that the predicted MTF may decrease as much as 17 times if

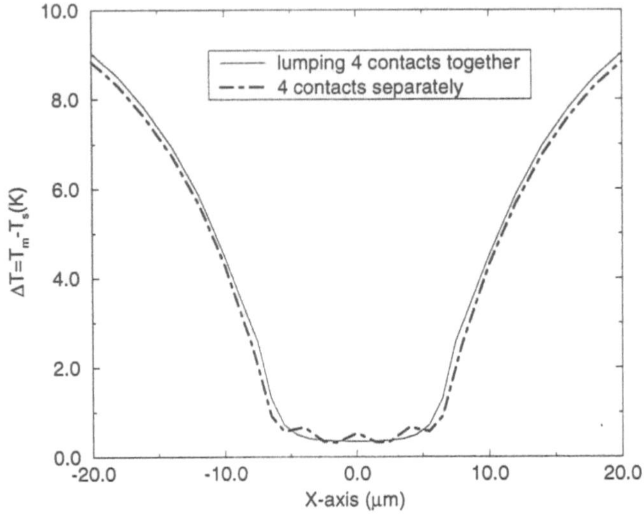

Figure 6.22. Simulation results of multiple contacts which are close to each other.

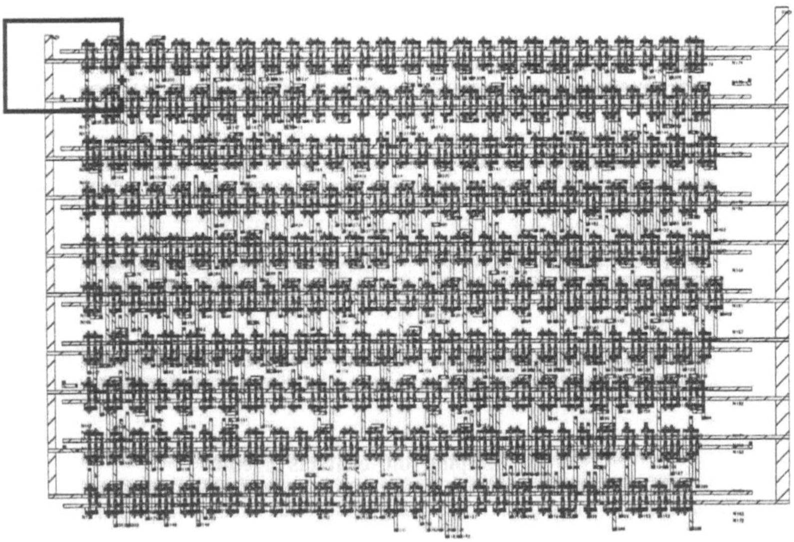

Figure 6.23. A layout of 10-bit negative adder.

the heating effects are considered. Therefore, the electrothermal analysis must be done prior to the electromigration diagnosis in order to pinpoint the true locations that are susceptible to electromigration problems.

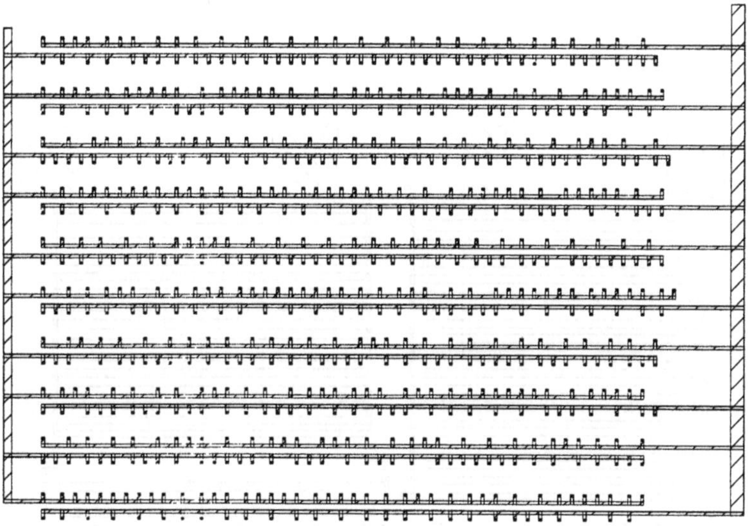

Figure 6.24. The power/ground bus layout of 10-bit negative adder.

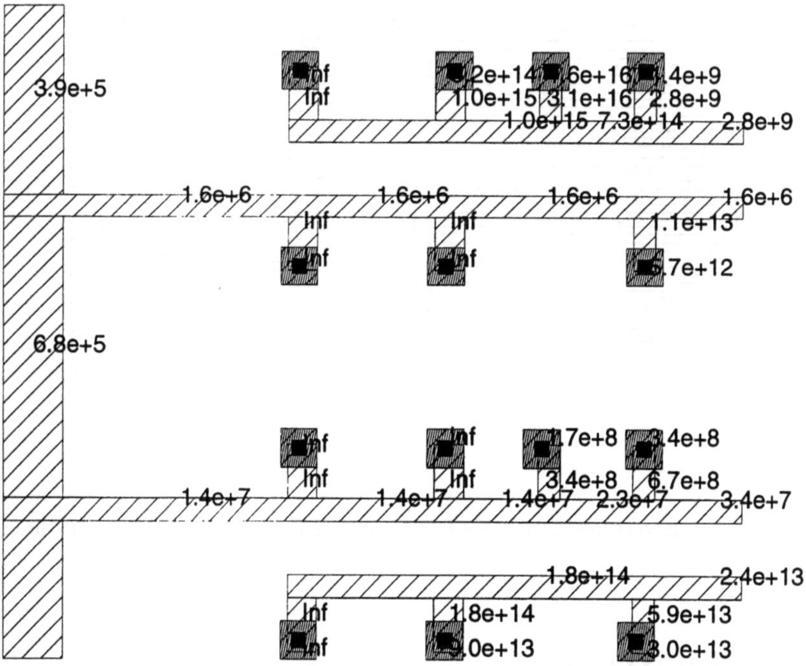

Figure 6.25. iTEM simulation result of the 10-bit negative adder. The number marked is the predicted electromigration MTF in hours.

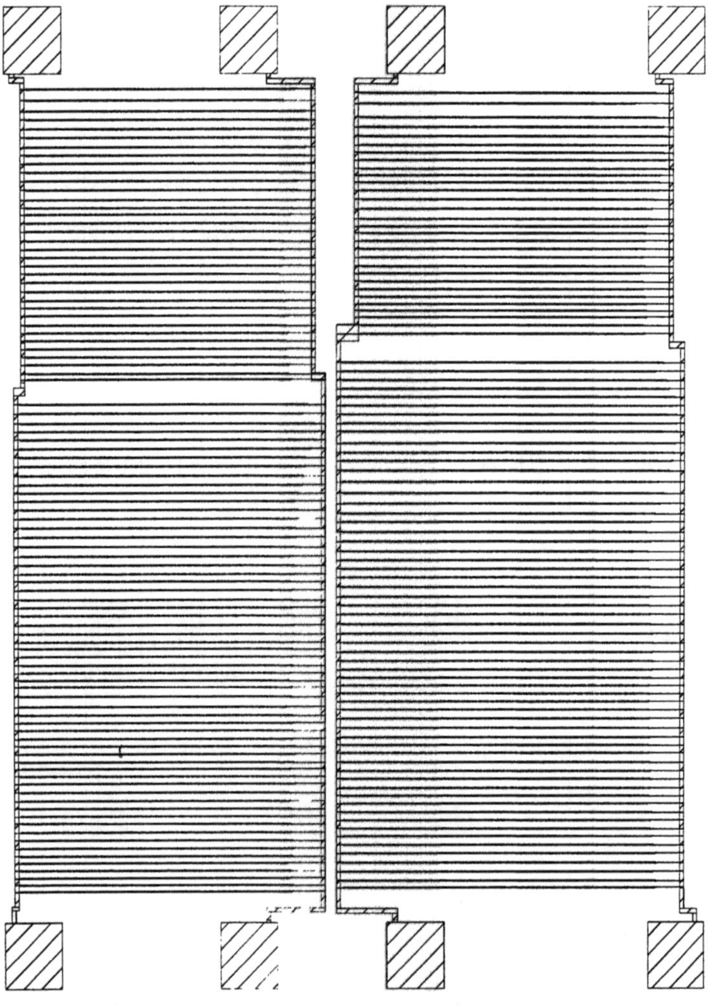

Figure 6.26. The power and ground bus layouts of the 2-D discrete cosine transformation chip.

6.5. SUMMARY

In this chapter, the temperature-dependent electromigration (EM) reliability diagnosis is discussed. Because the EM-induced mean time to failure is inversely proportional to the interconnect temperature, the temperature effect

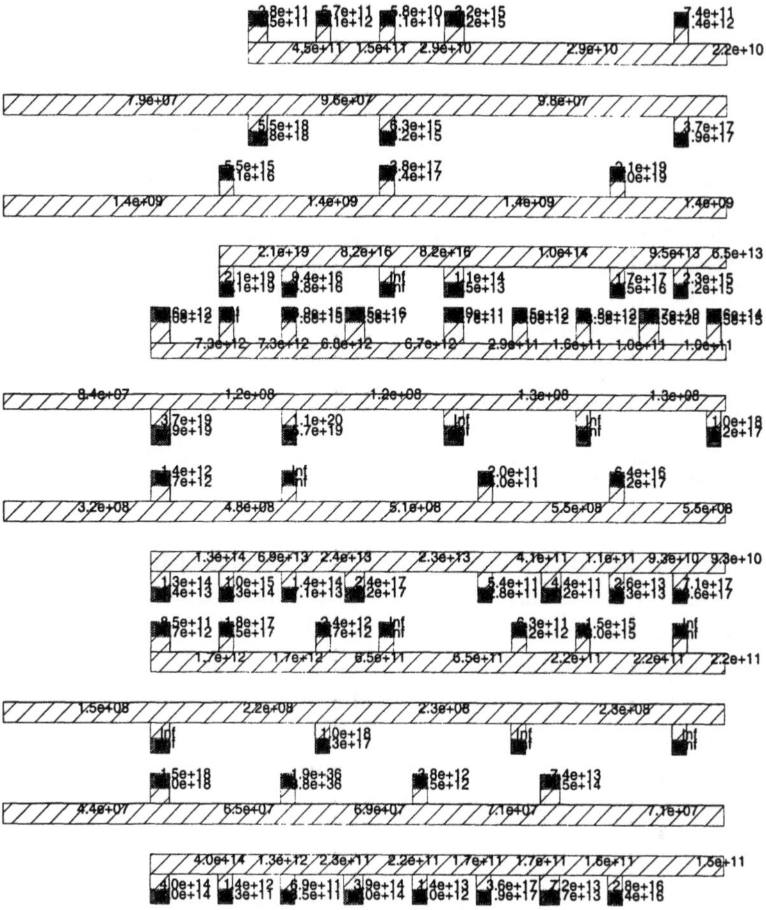

Figure 6.27. iTEM simulation result of the 2-D discrete cosine transformation chip. The number marked is the predicted electromigration MTF in hours.

is significant and must be taken into account in electromigration diagnosis in order to accurately predict the interconnect lifetime.

- The cause of electromigration phenomena is described.
- In addition to temperature, the electromigration lifetime is dependent on:
  - interconnect current density
  - interconnect current waveform
  - interconnect width
  - interconnect length

Table 6.1. Simulation results of iTEM.

Circuit		10-bit Adder	ALU and Control	16-bit Multiplier	2-D DCT
No. of Transistors		868	5842	10706	107832
Operation Freq.		300MHz	100MHz	100MHz	8MHz
Avg. Substrate Temperature		65.38°C	41.23°C	39.21°C	29.21°C
Avg. Metal Temperature		65.40°C	41.25°C	39.44°C	29.78°C
Peak Metal Temperature		67.57°C	44.19°C	69.55°C	79.95°C
CPU Time (Sec)		76	848	1177	≈ 12 hours
Estimated MTF(hours)	w/o thermal simulation (T=27°C)	6.8×10^6	5.84×10^6	5.29×10^5	4.92×10^6
	with thermal simulation	3.91×10^5	1.1×10^6	6.27×10^4	2.33×10^5

The ambient temperature is assumed 27°C.

For the current density, the inverse-current-square relationship is commonly observed. For the current waveform, the mean times to failure resulting from the unidirectional and bidirectional current stress modes are derived and compared. An average current model is used in both modes, while the healing effect is considered in the bidirectional mode. The interconnect lifetime dependence on its width is examined for the metal with the triple-point structure and the bamboo structure. The interconnect lifetime dependence on its length is modeled based on the series model with slight modification.

- The electromigration model for circuit simulation used in this book is given.
- An overview of existing electromigration diagnosis tools is provided.
- The temperature-dependent electromigration diagnosis tool, called iTEM, is introduced. This tool takes into account the interconnect temperature as one of its modeling parameters.
  - The interconnect temperature is estimated based on the substrate temperature, which can be found by thermal simulation introduced in Chapter 4.
  - Three assumptions are made in iTEM electromigration analysis.
  - An analytical model for estimating the interconnect temperature is derived. Although this model is simple and general accurate, it cannot accurately estimate the temperature near the heat sinks such as the metal contact (via) and the bond pad.

- To remedy the above accuracy problem, a lumped thermal model is derived for estimating the temperatures of the interconnect system. This model utilizes the thermal-electrical analogy, and the interconnect temperatures can be found by solving the node voltages of the thermal circuit. The thermal circuit is comprised of a constant current source to model the Joule heating, a voltage-dependent current source to model the resistivity increase caused by rising temperature, and thermal resistances. The analytical formulae for finding the thermal resistance of the L-shaped and T-shaped metals are provided.
- The layout partitioning and contact grouping strategies in the iTEM analysis flow are presented.
- Finally, iTEM simulation examples are given. It is shown that the mean time to failure of the interconnect is significantly reduced if the temperature effect is considered.

References

[1] J. R. Black, "Electromigration failure modes in aluminum metalization for semiconductor devices," *Proceedings of the IEEE*, vol. 57, pp. 1587–1594, Sept. 1969.

[2] K. Hinode, T. Furusawa, and Y. Homma, "Dependence of electromigration lifetime on the square of current density," in *Proceedings of the IEEE International Reliability Physics Symposium*, pp. 317–326, 1993.

[3] M. Sakimoto, T. Itoo, T. Fujii, H. Yamaguchi, and K. Eguchi, "Temperature measurement of Al metallization and the study of Black's model in high current density," in *Proceedings of the IEEE International Reliability Physics Symposium*, pp. 333–341, 1995.

[4] J. M. Towner and E. P. van de Ven, "Aluminum electromigration under pulsed DC conditions," in *Proceedings of the IEEE International Reliability Physics Symposium*, pp. 36–39, 1983.

[5] L. Brooke, "Pulsed current electromigration failure model," in *Proceedings of the IEEE International Reliability Physics Symposium*, pp. 136–139, 1987.

[6] J. A. Maiz, "Characterization of electromigration under bidirectional (BC) and pulsed unidirectional (PDC) currents," in *Proceedings of the IEEE International Reliability Physics Symposium*, pp. 220–228, 1988.

[7] J. J. Clement, "Vacancy supersaturation model for electromigration failure under DC and pulsed DC stress," *Journal of Applied Physics*, vol. 91, pp. 4264–4268, May 1992.

[8] J. Tao, N. Cheung, and C. Hu, "An electromigration failure model for interconnects under pulsed and bidirectional current stressing," *IEEE Transactions on Electron Devices*, vol. 41, pp. 539–545, Apr. 1994.

[9] B. K. Liew, N. Cheung, and C. Hu, "Projecting interconnect electromigration lifetime for arbitrary current waveforms," *IEEE Transactions on Electron Devices*, vol. 37, pp. 1343–1351, May 1990.

[10] J. Tao, K. Young, C. A. Pico, N. Cheung, and C. Hu, "Electromigration characteristics of Al/W via contact under unidirectional and bidirectional current conditions," in *Proceedings of the IEEE VLSI Multilevel Interconnection Conference*, pp. 390–392, 1991.

[11] J. Tao, N. Cheung, and C. Hu, "Metal electromigration damage healing under bidirectional current stress," *IEEE Electron Device Letters*, vol. 14, pp. 554–556, Dec. 1993.

[12] L. M. Ting, J. S. May, W. R. Hunter, and J. W. McPherson, "AC electromigration characterization and modeling of multilayered interconnects," in *Proceedings of the IEEE International Reliability Physics Symposium*, pp. 311–316, 1993.

[13] J. Tao, N. W. Cheung, and C. Hu, "Modeling electromigration lifetime under bidirectional current stress," *IEEE Electron Device Letters*, pp. 476–478, Nov. 1995.

[14] T. Kwok, "Effect of metal line geometry on electromigration lifetime in Al-Cu submicron interconnects," in *Proceedings of the IEEE International Reliability Physics Symposium*, pp. 185–191, 1988.

[15] B. N. Agarwala, M. J. Attardo, and A. P. Ingraham, "Dependence of electromigration-induced failure time on length and width of aluminum thin-film conductors," *Journal of Applied Physics*, vol. 41, pp. 3954–3960, Oct. 1970.

[16] T. Kwok, J. Finnegan, and D. Johnson, "Effect of linelength and bend structure on electromigration lifetime in Al-Cu submicron interconnects," in *Proceedings of the IEEE VLSI Multilevel Interconnection Conference*, pp. 436–445, 1988.

[17] T. Nogami, S. Oka, K. Naganuma, T. Nakata, C. Maeda, and O. Haida, "Electromigration lifetime as a function of line length or step number,"

in *Proceedings of the IEEE International Reliability Physics Symposium*, pp. 366–372, 1992.

[18] D. F. Frost and K. F. Poole, "A method for predicting VLSI-device reliability using series models for failure mechanisms," *IEEE Transactions on Reliability*, vol. R-36, pp. 234–242, June 1987.

[19] J. E. Hall, D. E. Hocevar, P. Yang, and M. J. McGraw, "SPIDER– A CAD system for modeling VLSI metallization patterns," *IEEE Transactions on Computer-Aided Design of Integrated Circuits and Systems*, vol. CAD-36, pp. 1023–1031, Nov. 1987.

[20] L. W. Nagel, *SPICE2: A Computer Program to Simulate Semiconductor Circuits.* PhD thesis, Dept. of Electrical Engineering, University of California at Berkeley, 1975.

[21] D. F. Frost and K. F. Poole, "RELIANT: A reliability analysis tool for VLSI interconnects," *IEEE Journal of Solid-State Circuits*, vol. 24, pp. 458–462, Apr. 1989.

[22] D. A. Haeussler and K. F. Poole, "CURRANT: A current prediction software tool using a switch-level simulator," in *IEEE Southeastcon '89 Proceedings*, pp. 946–948, 1989.

[23] R. H. Tu, E. Rosenbaum, W. Y. Chan, C. C. Li, E. Minami, K. Quader, P. K. Ko, and C. Hu, "Berkeley reliability tools-BERT," *IEEE Transactions on Computer-Aided Design of Integrated Circuits and Systems*, vol. 12, pp. 1524–1534, Oct. 1993.

[24] T. S. Hohol and L. A. Glasser, "RELIC: A reliability simulator for integrated circuits," in *Proceedings of the ACM/IEEE International Conference on Computer-Aided Design*, pp. 517–520, Nov. 1986.

[25] B. J. Sheu, W.-J. Hsu, and B. W. Lee, "An integrated circuit reliability simulator – RELY," *IEEE Journal of Solid-State Circuits*, vol. 24, pp. 473–477, Apr. 1989.

[26] F. N. Najm, R. Burch, P. Yang, and I. N. Hajj, "CREST - a current estimation for CMOS circuits," in *Proceedings of the ACM/IEEE International Conference on Computer-Aided Design*, pp. 204–207, 1988.

[27] F. N. Najm, R. Burch, P. Yang, and I. N. Hajj, "Probabilistic simulation for reliability analysis of CMOS VLSI circuits," *IEEE Transactions on Computer-Aided Design of Integrated Circuits and Systems*, vol. 9, pp. 439–450, Apr. 1990.

[28] F. N. Najm, I. N. Hajj, and P. Yang, "An extension of probabilistic simulation for reliability analysis of CMOS VLSI circuits," *IEEE Transactions on Computer-Aided Design of Integrated Circuits and Systems*, vol. 10, pp. 1372–1381, Nov. 1991.

[29] C. C. Teng, Y. K. Cheng, E. Rosenbaum, and S. M. Kang, "Hierarchical electromigration reliability diagnosis for VLSI interconnects," in *Proceedings of the ACM/IEEE Design Automation Conference*, pp. 752–757, June 1996.

[30] C. C. Teng, Y. K. Cheng, E. Rosenbaum, and S. M. Kang, "iTEM: A new electromigration (EM) reliability diagnosis tool using electrothermal timing simulation," in *Proceedings of the IEEE International Reliability Physics Symposium*, pp. 172–179, 1996.

[31] Y. K. Cheng, P. Raha., C. C. Teng, E. Rosenbaum, and S. M. Kang, "ILLIADS-T: An electrothermal timing simulator for temperature-sensitive reliability diagnosis of CMOS VLSI chips," *IEEE Transactions on Computer-Aided Design of Integrated Circuits and Systems*, vol. 17, pp. 668–681, Aug. 1998.

[32] R. M. Iimura, "iCHARM: Hierarchical CMOS circuit extraction with power bus extraction," Master's thesis, Dept. of Electrical and Computer Engineering, University of Illinois at Urbana-Champaign, 1990.

[33] H. A. Schafft, "Thermal analysis of electromigration test structures," *IEEE Transactions on Electron Devices*, pp. 664–672, Mar. 1987.

[34] A. A. Bilotti, "Static temperature distribution in IC chips with isothermal heat source," *IEEE Transactions on Electron Devices*, pp. 217–226, Mar. 1974.

[35] H. Katto, M. Harada, and Y. Higuchi, "Wafer-level JRAMP and J-CONSTANT electromigration testing of conventional and SWEAT patterns assisted by a thermal and electrical simulator," in *Proceedings of the IEEE International Reliability Physics Symposium*, pp. 85–88, 1991.

[36] *THUNDER User's Manual.* SILVACO Data Systems, 1993.

[37] S. L. Su, *Extraction of MOS VLSI Circuits Models Including Critical Interconnect Parasitics.* PhD thesis, Dept. of Electrical and Computer Engineering, University of Illinois at Urbana-Champaign, 1987.

[38] S. L. Su, V. B. Rao, and T. N. Trick, "HPEX: A hierarchical parasitic circuit extractor," in *Proceedings of the ACM/IEEE Design Automation Conference*, pp. 566–569, 1987.

[39] Y. C. Lee, H. T. Ghaffari, and J. M. Segelken, "Internal thermal resistance of a multi-chip packaging design for VLSI based system," *IEEE Transactions on Components, Hybrids and Manufacturing Technology*, pp. 163–169, June 1989.

[40] J. W. Stroming, *VHDL Synthesis of the Two-Dimensional Discrete Cosine Transform*. PhD thesis, Dept. of Electrical and Computer Engineering, University of Illinois at Urbana-Champaign, 1995.

Chapter 7

TEMPERATURE-DRIVEN CELL PLACEMENT

7.1. INTRODUCTION

The smaller feature size, higher packing density and rising power consumption have led to dramatic temperature increase in modern VLSI circuits. Moreover, the cross-chip temperature differential larger than tens of degrees has been commonly observed. Because the circuit delay and many IC reliability problems are strongly temperature dependent, the hot spots often become the performance and reliability bottlenecks and create serious design constraints as illustrated in previous chapters. Consequently, the capability to assess and optimize the thermal quality throughout the VLSI design process is critically important. Since the thermal (temperature) distribution profile of a design is largely determined by the cell locations, cell placement is the natural starting point of a temperature-aware design flow. In this chapter, it will be shown that careful cell placement at the physical design stage can help improve the thermal distribution of the design, adding little or only minor overhead to conventional design objectives such as area and delay.

7.2. OVERVIEW

The majority of the previous studies on the thermal placement problem were mostly conducted in the context of placing chips for printed circuit boards (PCBs) and multi-chip modules (MCMs) [1]-[4]. Due to the differences in boundary conditions and problem granularity, these results are not directly applicable at the cell level.

A more relevant study can be found in [5], where the authors proposed a generic force-directed placement algorithm that can potentially incorporate the power distribution of a placement as one of the placement considerations. In [6] the authors modeled the thermal placement problem as a matrix synthesis

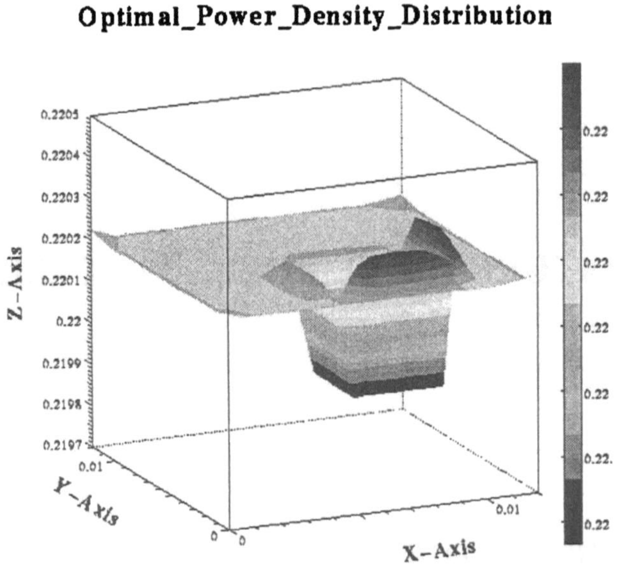

Figure 7.1(a). Optimal heat distribution for a design with a core size 12mm × 12mm. The power density of the fixed cell near the lower-right corner of the layout is lower than the chip average.

problem where the temperature distribution is optimized by uniform distribution of heat sources (gates) on the chip.

There is one common issue that the above two papers did not address: Due to the finite thermal conductivity of the packaging components and the presence of the hard-placed cells, uniform heat distribution *does not* lead to uniform temperature profile. Therefore, if the thermal placement problem is formulated as to optimally distribute the heat sources, then the optimal heat distribution under the given thermal boundary conditions and design specifications (i.e. the location and power dissipation of any hard-placed cell) must be found first. Figures 7.1(a) and 7.1(b) illustrates this concept. A particular design with one fixed cell and with a total power of 30 Watts is used in this example. Figure 7.1(a) shows the optimal heat distribution of the design in order to produce the temperature profile in Fig. 7.1(b) that is uniformly flat outside the fixed cell. It can be seen that the optimal heat distribution is not uniform due to the effects of boundary conditions and the existence of hard-placed cells.

Another limiting factor of the approach taken in [6] is the assumption of constant gate power dissipation. The power dissipated by individual cells is affected by the load capacitances, which do not remain constant during the placement as the locations of other cells change. As a result, the cell power dissipation in the final placement may be significantly different from the cell power

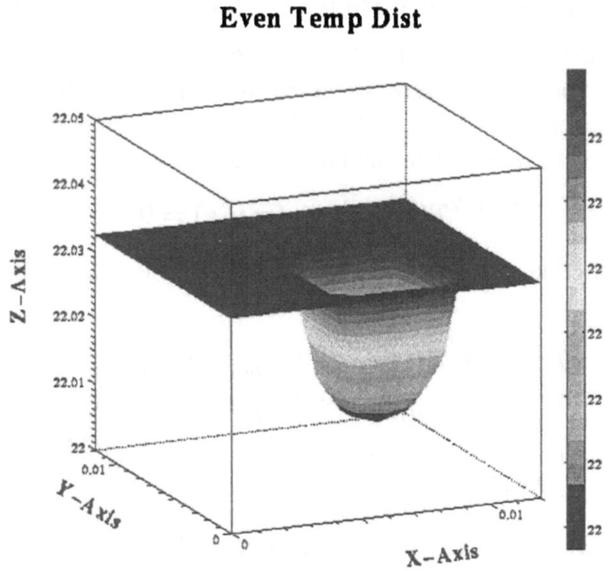

Figure 7.1(b). Optimal temperature distribution resulting from the heat distribution in Fig. 7.1(a).

calculated before placement. It is not clear how the matrix synthesis algorithm can adapt to on-the-fly power estimation required for thermal placement at the gate level.

In [7], the above limiting factors are addressed: The authors proposed a method for standard cell placement that aims to reduce the number of hot spots while optimizing traditional design metrics such as area and wire length. By using the superposition principle and the concept of transfer thermal resistance, the temperature distribution constraint is converted to its corresponding power distribution constraint under arbitrary boundary conditions. The power distribution constraint is then gradually tightened during the placement process by simulated annealing to produce improved temperature profile. Later in [8], an approach similar to [7] was proposed for macrocell thermal placement. A new thermal penalty term is added to the overall cost function during the modified simulated annealing process.

The thermal simulator in [7] and [8] calculates the steady-state substrate temperature based on the finite-difference method as described in chapter 4. Enhancement is made in their work to reduce the matrix size during numerical simulation by deriving a compact substrate thermal model. The remainder of this chapter will begin with the discussion of this model. Two thermal placement algorithms proposed in [7] and [8] will also be examined.

7.3. SUBSTRATE TEMPERATURE CALCULATION

The heat diffusion equation and its discretized form used in the numerical finite-difference formulation were discussed in Chapter 4. If the thermal conductivity k is uniform (i.e., independent of position and temperature), the heat diffusion equation at steady state can be expressed as

$$k \cdot \nabla^2 T(x,y,z) + g(x,y,z) = 0 \tag{7.1}$$

which is a linear equation. By combining Eq. (4.28), Eq. (4.29), and Eq. (7.1), we have

$$\begin{aligned}
&(T_{i,j,k} - T_{i+1,j,k})G1 + (T_{i,j,k} - T_{i-1,j,k})G4 + \\
&(T_{i,j,k} - T_{i,j+1,k})G2 + (T_{i,j,k} - T_{i,j-1,k})G5 + \\
&(T_{i,j,k} - T_{i,j,k+1})G3 + (T_{i,j,k} - T_{i,j,k-1})G6 \\
&= \Delta g(x,y,z)
\end{aligned} \tag{7.2}$$

where $\Delta g(x,y,z)$ is the discretized heat source, and all of the time-dependent terms are omitted for steady state. Equivalently, Eq. (7.2) can be written as

$$\begin{aligned}
&(T_N - T_{x1})g_{N,x1} + (T_N - T_{x2})g_{N,x2} + (T_N - T_{y1})g_{N,y1} + \\
&(T_N - T_{y2})g_{N,y2} + (T_N - T_{z1})g_{N,z1} + (T_N - T_{z2})g_{N,z2} \\
&= \Delta g(x,y,z).
\end{aligned} \tag{7.3}$$

By exploiting the thermal-electrical analogy, T_N can be shown to be the node voltage (temperature) at node $N = (i,j,k)$, $g_{i,j}$ be the (thermal) conductance between node i and node j, and $\Delta g(\cdot)$ be the injection current at node N. Equation (7.3) is in fact the KCL equation for node N. Writing the same equation for every node i in the substrate mesh and compiling these equations into a matrix form, we have

$$\mathbf{Gv} = \begin{bmatrix} g_{11} & g_{12} & \cdots & g_{1n} \\ g_{21} & g_{22} & \cdots & g_{2n} \\ \vdots & \vdots & \ddots & \vdots \\ g_{n1} & g_{n2} & \cdots & g_{nn} \end{bmatrix} \begin{bmatrix} v_1 \\ v_2 \\ \vdots \\ v_n \end{bmatrix} = \begin{bmatrix} i_1 \\ i_2 \\ \vdots \\ i_n \end{bmatrix} = \mathbf{i} \tag{7.4}$$

with $\mathbf{v}$ denoting the node temperature, and $\mathbf{i}$ the power dissipation at each node. These notations will be followed henceforth. Note that the matrix $\mathbf{G}$ in Eq. (7.4) is symmetric and positive definite.

The computational cost associated with solving the nodal matrix equation in Eq. (7.4) is too expensive to be used directly in an iterative placement algorithm for temperature calculation and optimization. Consequently, for temperature-aware physical design, either a more efficient temperature calculation method or an alternative approach is needed. In the following section it will be shown that this can be achieved by using a more compact substrate thermal model.

7.4. COMPACT SUBSTRATE THERMAL MODELING

In this section, two methods are discussed to derive the compact substrate thermal model for thermal placement. The first method employs the superposition principle to construct a *transfer thermal resistance matrix*, and is independent of the methods used to obtain the substrate temperature values. The superposition principle applies here because Eq. (7.1) is linear. The second method involves only direct manipulation of the nodal matrix equation, and is applicable when the user chooses a numerical method such as the finite-difference method that uses such a matrix equation for temperature calculation.

7.4.1 TRANSFER THERMAL RESISTANCE MATRIX

Assume that the substrate surface is discretized into a collection of m points where movable cells can reside. These points also act as the temperature monitor points. The temperature values at these m locations can be found by using a numerical method, or by experiments. By the *superposition principle*, the temperature values at these points are simply the sum of the separate temperature profiles created by individual heat sources in the system:

$$\begin{aligned} \mathbf{T}_{total} &= \sum \mathbf{T}_{movable\_cells} + \sum \mathbf{T}_{fixed\_cells} + \mathbf{T}_{ambient} \\ &= \sum \mathbf{T}_{movable\_cells} + \mathbf{T}_{BC} \end{aligned} \tag{7.5}$$

where $\mathbf{T}_{movable\_cells}$, $\mathbf{T}_{fixed\_cells}$ and $\mathbf{T}_{ambient}$ are the temperature profiles at the m monitor points caused by movable cells, hard-placed cells and the ambient individually. Here, $\mathbf{T}_{BC}$ stands for the temperature set by the ambient and fixed cells. Note that vectors such as $\mathbf{T}_{BC}$ are shown in bold type. Applying the superposition principle further to the monitor points on the substrate surface that are to be covered by movable cells, we have:

$$\mathbf{T}_{total} = \sum_{i=1}^{m} \mathbf{T}_i + \mathbf{T}_{BC} \tag{7.6}$$

where $\mathbf{T}_i$ is the temperature profile caused by the power located at monitor point i alone.

Let us define the *transfer thermal resistance* R^t_{ij} of point i with respect to point j as the rise in temperature at point i due to one unit of power dissipated at point j:

$$R^t_{ij} = \frac{\Delta t_{ij}}{\Delta p_j}. \tag{7.7}$$

We can define a R^t_{ij} for every grid point pair (i, j) among the m surface monitor points, and represent these resistances in the matrix form

$$\mathbf{R}^t = \begin{bmatrix} R^t_{11} & R^t_{12} & \cdots & R^t_{1m} \\ R^t_{21} & R^t_{22} & \cdots & R^t_{2m} \\ \vdots & \vdots & \ddots & \vdots \\ R^t_{m1} & R^t_{m2} & \cdots & R^t_{mm} \end{bmatrix}. \tag{7.8}$$

The matrix $\mathbf{R}^t$ is the *transfer thermal resistance matrix*. For any cell power distribution $P = [P_1, P_2, \cdots, P_m]$ on the monitor grid points, $\mathbf{T}_{movable\_cells}$ can be calculated by direct matrix multiplication:

$$\begin{aligned} \sum \mathbf{T}_{movable\_cells} &= \begin{bmatrix} T_1 \\ T_2 \\ \vdots \\ T_m \end{bmatrix}_{movable\_cells} \\ &= \begin{bmatrix} R^t_{11} & R^t_{12} & \cdots & R^t_{1m} \\ R^t_{21} & R^t_{22} & \cdots & R^t_{2m} \\ \vdots & \vdots & \ddots & \vdots \\ R^t_{m1} & R^t_{m2} & \cdots & R^t_{mm} \end{bmatrix} \begin{bmatrix} P_1 \\ P_2 \\ \vdots \\ P_m \end{bmatrix}. \end{aligned} \tag{7.9}$$

The jth column in the transfer thermal resistance matrix, $[R^t_{j1}, R^t_{j2}, \cdots, R^t_{jm}]^T$, is simply the temperature at points $1, 2, \cdots, m$ of the monitored grid points due to one Watt of power dissipated at point j, and can be calculated by using any temperature calculation method or measured experimentally. Combining Eq. (7.5) and Eq. (7.9), we have

$$\mathbf{T}_{total} - \mathbf{T}_{BC} = \begin{bmatrix} R^t_{11} & R^t_{12} & \cdots & R^t_{1m} \\ R^t_{21} & R^t_{22} & \cdots & R^t_{2m} \\ \vdots & \vdots & \ddots & \vdots \\ R^t_{m1} & R^t_{m2} & \cdots & R^t_{mm} \end{bmatrix} \begin{bmatrix} P_1 \\ P_2 \\ \vdots \\ P_m \end{bmatrix} \tag{7.10}$$

where $\mathbf{T}_{BC}$ can also be found by using any computational or experimental method.

Once $\mathbf{R}^t$ and $\mathbf{T}_{BC}$ are obtained, the temperature profile due to any power distribution can be calculated by direct matrix multiplication. Alternatively, for any desired thermal distribution, one can determine the corresponding power distribution by treating $[P_1, P_2, \cdots, P_m]^T$ as unknowns and solving for Eq. (7.10). As an example, let us define the optimal substrate temperature profile as one that is perfectly uniform, and the optimal power distribution as one that creates such an optimal temperature profile. Then for the given set of boundary conditions and total power dissipation, the optimal power distribution and the corresponding uniform substrate temperature can be obtained by solving the following matrix equation:

$$\begin{bmatrix} R_{11}^t & R_{12}^t & \cdots & R_{1m}^t & -1 \\ R_{21}^t & R_{22}^t & \cdots & R_{2m}^t & -1 \\ \vdots & \vdots & \ddots & \vdots & \vdots \\ R_{m1}^t & R_{m2}^t & \cdots & R_{mm}^t & -1 \\ 1 & 1 & \cdots & 1 & 0 \end{bmatrix} \times \begin{bmatrix} P_1 \\ P_2 \\ \vdots \\ P_m \\ T_s \end{bmatrix} = \begin{bmatrix} -T_1^{bc} \\ -T_1^{bc} \\ \vdots \\ -T_m^{bc} \\ P_{total} \end{bmatrix} \tag{7.11}$$

where $\mathbf{T}_{BC} = [T_1^{bc}, T_2^{bc}, \cdots, T_m^{bc}]^T$, P_{total} is the sum of the power dissipated by all movable cells, $\mathbf{P} = [P_1, P_2, \cdots, P_m]$ is the optimal power distribution with $\sum_{i=1}^{m} P_i = P_{total}$, and T_s is the optimal temperature under the given total power and boundary conditions. This is how the optimal power distribution map was generated in Fig. 7.1(a).

7.4.2 ADMITTANCE MATRIX REDUCTION

The transfer thermal resistance matrix can also be obtained through direct reduction of the admittance matrix that is constructed by numerical methods such as the finite-difference method for temperature calculation. Assuming the 3-D substrate mesh has $m+1$ port nodes (including the thermal ground) and n internal nodes, the thermal conductance matrix $\mathbf{G}$ in Eq. (7.4) has $m+n$ rows and columns. If the nodes are reordered such that the first m rows correspond to the port nodes and the final n rows to the internal nodes, Eq. (7.4) can be rewritten as

$$\begin{bmatrix} \mathbf{G}_P & \mathbf{G}_C^T \\ \mathbf{G}_C & \mathbf{G}_I \end{bmatrix} \begin{bmatrix} \mathbf{v}_P \\ \mathbf{v}_I \end{bmatrix} = \begin{bmatrix} \mathbf{i}_P \\ 0 \end{bmatrix} \tag{7.12}$$

where $\mathbf{v}_P$ and $\mathbf{v}_I$ represent the m port temperatures and n internal node temperatures, respectively, and $\mathbf{i}_P$ denotes the power dissipation at the port nodes.

The dimensions of the submatrices in Eq. (7.12) are $m \times m$ for $\mathbf{G}_P$, $n \times m$ for $\mathbf{G}_C$, and $n \times n$ for $\mathbf{G}_I$. Note that the internal node part of $\mathbf{i}$ is zero. It is because there is no heat dissipated at these nodes in a chip (i.e., heat sources are normally distributed on the top surface). If the multiport admittance is defined as $\mathbf{Y} \equiv \mathbf{i}_P/\mathbf{v}_P$, and $\mathbf{v}_I$ is eliminated in Eq. (7.12), we have

$$\mathbf{Y} = \mathbf{G}_P - \mathbf{G}_C^T \cdot \mathbf{G}_I{}^{-1} \cdot \mathbf{G}_C. \tag{7.13}$$

The power dissipations at the port nodes and their temperatures are thus related by a simple matrix equation:

$$(\mathbf{G}_P - \mathbf{G}_C^T \mathbf{G}_I{}^{-1} \mathbf{G}_C)\mathbf{v}_P = \mathbf{Y}\mathbf{v}_P = \mathbf{i}_P. \tag{7.14}$$

All internal nodes in Eq. (7.14) have been entirely eliminated, which results in a much smaller admittance matrix $\mathbf{Y}$ of dimension $m \times m$, compared with the original $\mathbf{G}$ of size $(m+n) \times (m+n)$. The reduced network has exactly the same input/output characteristics as the original network; no additional error is introduced by the reduction. Because $\mathbf{G}_I$ is symmetric and positive definite, the Cholesky factorization can be applied to compute the inverse of $\mathbf{G}_I$, which is more efficient than the LU factorization.

The matrix $\mathbf{Y}$ is in fact the inverse of the transfer thermal resistance matrix derived in the previous section:

$$\mathbf{Y} \cdot \mathbf{R}^t = \mathbf{I}. \tag{7.15}$$

The resulting $\mathbf{Y}$ is usually very dense, meaning that the reduced thermal resistive network is strongly connected.

7.4.3 RUNTIME EFFICIENCY OF COMPACT THERMAL MODELING

The following complexity analysis is based on the assumption that, the same network admittance matrix constructed by the numerical method (such as the finite-difference method) is used for both compact model derivation methods described in Sections 7.4.1 and 7.4.2. Again, assuming there are $m+1$ port nodes (including the thermal ground) and n internal nodes in the original thermal network.

The first method, described in Section 7.4.1, involves decomposing the admittance matrix of size $(m+n) \times (m+n)$ once, whose runtime complexity is $O((m+n)^3)$ if LU or Cholesky factorization is used. In addition, it requires forward/backward substitutions m times, whose complexity is $O(m(m+n)^2)$. Therefore the total complexity of the first method is $O((m+n)^3) + O(m(m+n)^2) \approx O((m+n)^3)$.

The runtime complexity of the second method, introduced in Section 7.4.2, can be obtained by analyzing the complexity of Eq. (7.14). Eq. (7.14) includes inverting the matrix $\mathbf{G}_I$ of dimension $n \times n$, two matrix multiplications, and a matrix subtraction. The complexity of inverting $\mathbf{G}_I$ is $O(n^3)$; the complexity of the multiplications is $O(mn^2)$; and the complexity of the subtraction is $O(m^2)$. The total complexity of the second method is thus $O(n^3)+O(mn^2)+O(m^2) \approx O(n^3)$ if $m \ll n$, or $O(n^3) + O(mn^2)$ otherwise.

Since $m < n$ almost always holds in practice, it can be easily seen that the second method is more efficient. The first method should be used only when experimental results are readily available and can be used to construct the transfer thermal resistance matrix $\mathbf{R}^t$ without further computation, or if the temperature calculation tools at hand do not utilize the mesh network as the substrate thermal model (such as those based on the Green's function solutions).

It is important to point out that the simple discussion carried out above does not consider the performance improvement of employing advanced sparse matrix solving techniques. More careful and complicated analysis is required if such techniques are to be used.

7.5. THERMAL PLACEMENT ALGORITHMS

The block diagram of the thermal placement algorithm based on the compact thermal modeling is shown in Fig. 7.2. The whole process consists of three main steps:

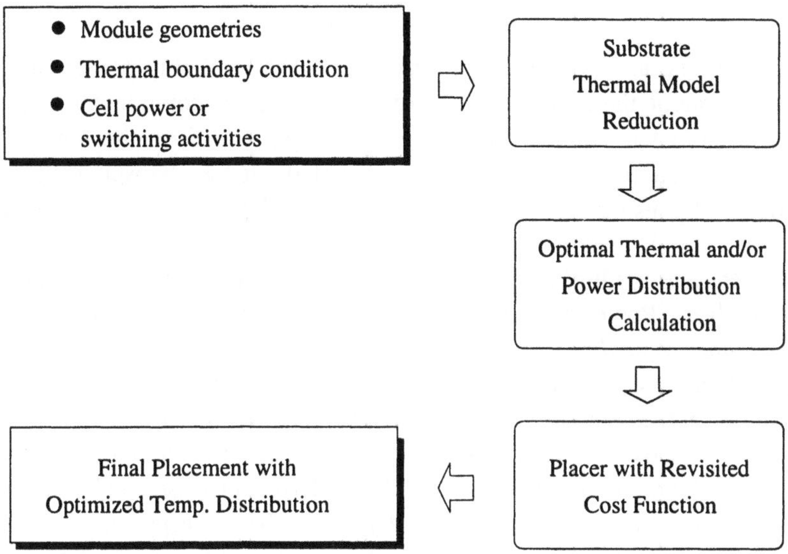

Figure 7.2. Block diagram of the thermal placement algorithm.

1. *substrate thermal model derivation*
2. *thermal objective construction*
3. *placement with thermal objective optimization*

For reasons that will become clear, standard cell and macro cell designs require different thermal objectives and strategies for thermal distribution optimization. In the following these steps will be discussed in detail for both design styles.

7.5.1 STANDARD CELL THERMAL PLACEMENT

The first step in standard cell thermal placement is to construct the matrix **Y**. As described in Section. 7.4.2, the matrix **Y** is reduced from the network admittance matrix **G** that is built by using the finite-difference method. Next, the matrix **Y** is used to convert the user specified thermal (temperature) distribution objective into the corresponding power distribution objective. The specified thermal distribution objective can be either a uniform or a non-uniform thermal distribution, which is based upon the locations of the temperature-sensitive components such as the analog or the clock generation modules. In the following discussion, it is assumed that a uniform thermal distribution is used as the objective, and this objective is determined from the average substrate temperature.

With the estimated total chip power, the average substrate temperature is first calculated by using

$$T_{avg} = T_a + P_{chip} \times R^{th} \tag{7.16}$$

where T_a is the ambient temperature, P_{chip} is the total chip power, and R^{th} is the equivalent thermal resistance of the packaging components. Then a user-adjustable temperature *slack* is added to T_{avg} to form the temperature objective. This temperature objective is converted into the corresponding power distribution by using Eq. (7.14). The calculated power constraint $\mathbf{P}_{obj} = [P_1, P_2, \cdots, P_m]^T$ becomes the *final power budget* at the m port grid points. Finally, $\mathbf{P}_{obj}$ is multiplied by a user-specified factor α to get the *starting power budget*. Power budgets are used to prune the cell movements during the *simulated annealing* process. Specifically, a cell movement changes the lengths of the nets that the cell is connected to, thereby changing the power dissipation on the cells directly connected to the moving cell.

For any proposed cell movement, the power dissipation changes at all m monitor grid points are first recalculated. If the proposed cell movement only increases the power at those grid points where the power dissipations do not exceed the current budget, then the movement passes the budget test. Conversely, for the grid points where the power already exceed the budget, if the proposed movement further increases their power dissipation, then this movement is immediately rejected. The power budget is gradually tightened from the starting value $\alpha\mathbf{P}_{obj}$ to the final value $\mathbf{P}_{obj}$ during the placement process according to the cooling schedule. The revised simulated annealing algorithm is shown in Fig. 7.3.

There are a few issues worth further discussion:

1. The estimated total power in Eq. (7.16), from which the average substrate temperature is estimated, can be provided by the user. Or alternatively, a few number of random placements can be generated in order to obtain a quick estimate of the total power dissipation by using the following equation:

$$P_{total} = \sum_{i=1}^{\#cells} 0.5 \cdot f_{clk} \cdot V_{DD}^2 \cdot C_{load}(i) \cdot S_{rate}(i) \tag{7.17}$$

   where f_{clk} is the clock frequency, V_{DD} is the power supply voltage, $C_{load}(i)$ is the load capacitance of cell i, and $S_{rate}(i)$ is the rate of switching activity at the output pin of cell i.

```
Algorithm SIMULATED ANNEALING
   do
     do
        Generate_Movement();
        Recompute_Power_Dist();
        reject=Check_Power_Budget();
        if (reject) continue;
        ΔC = Compute_Cost_Change();
        Accept(ΔC, T);
     until in equilibrium;
     Reduce(T);
     Reduce_Budget(T);
   until cost cannot be further reduced;
End SIMULATED ANNEALING
```

Figure 7.3. Revised simulated annealing algorithm for standard cell thermal placement.

2. Given the final power budget $\mathbf{P}_{obj}$, one might be tempted to adopt a penalty-based approach for thermal optimization by adding an additional thermal penalty term to the cost function for the simulated annealing engine to optimize. However, unlike typical penalties such as cell overlaps or timing violations, thermal penalty usually cannot be completely eliminated by placement alone. Thus the presence of thermal penalty usually results in suboptimal placements. Depending on the weightings of different objectives, it can be shown from experiments that the penalty-based approach could result in up to 50% increase in total wirelength and up to 20% increase in area [7], which underperforms the above constraint-based method. Tradeoffs between the traditional and the thermal objectives will be demonstrated later in the simulation results section.

3. It is not necessary to add an equal amount of temperature slack to the average substrate temperature to form the thermal distribution constraint. For instance, if the design contains certain temperature sensitive subcircuits, it is beneficial to enforce a more rigid temperature constraint locally, while relaxing the constraint in non-critical regions.

4. The thermal distribution calculation is only as accurate as the power estimation. Accurate power estimation is a research topic that is being actively studied, and a general overview has been given in Chapter 2. The thermal placement algorithm introduced above is quite general and can be used with any power estimation technique. For instance, Eq. (7.17) can be replaced with a more accurate power measure.

5. The above constraint-based placement method can also be applied to circuits with multiple power dissipation patterns, such as the sequential circuits or the microprocessors with gated clocks. One only needs to apply the power budget test (i.e. `Check_Power_Budget()`) in the algorithm to all possible power dissipation scenarios.

7.5.2 MACROCELL THERMAL PLACEMENT

The reasons why the macrocell thermal placement demands a different implementation strategy from the standard cell thermal placement can be best understood by considering the sources of hot spots in a design. There are basically two such sources: One is the fast-switching, high power-consumption cells and the other is the thermal coupling between nearby hot cells.

Although the power dissipation of standard cells are affected by their wire load and thus their relative cell locations, it is safe to assume that for macrocells the power remain relatively constant regardless of the placement. As a result, a flatter power distribution cannot be enforced simply by moving cells around, as what is done in the case of standard cell design. The only way to improve the thermal distribution quality is by reducing the degree of *thermal coupling* between cells. Thus it is important for the thermal placement objective to capture the coupling effect. The most straightforward way to do so is to use the actual substrate surface temperature. To avoid the pitfall of lengthy simulation during placement, the compact substrate thermal model can be used to calculate the temperature efficiently.

Again the placement process begins with the construction of the $\mathbf{Y}$ matrix. It is then inverted to form the transfer thermal resistance matrix $\mathbf{R}^t$. Any desired thermal distribution can still be specified as the objective. During simulated annealing, the thermal profile is updated incrementally as follows. Suppose at a particular iteration the annealer moves cell a with heat dissipation P_a from grid point i to point j. The following new vector can be formed:

$$\mathbf{P}' = [0, \ldots, 0, -P_a, 0, \ldots, 0, P_a, 0, \ldots, 0]^T$$

with the entry $-P_a$ at location i and P_a at location j, respectively. The incremental temperature profile change is then calculated by $\mathbf{T}' = \mathbf{R}^t \times \mathbf{P}'$. The vector $\mathbf{T}'$ can subsequently be added back to the original temperature profile to obtain the final profile. Thermal penalty is calculated according to the degree of discrepancy between the new profile and the desired one.

It is important for the thermal penalty to discourage uneven temperature profile, as well as to reduce the maximum on-chip temperature. Inadequately constructed thermal penalty might in some cases result in very hot spots within very small areas. For the implementation in [8], the following formula is used

```
Algorithm SIMULATED ANNEALING
    do
      do
          Generate_Movement();
          Update_Temp_Dist();
          ΔC = Compute_Cost_Change();
          Accept(ΔC, T);
      until in equilibrium;
      Reduce(T);
    until cost cannot be further reduced;
End SIMULATED ANNEALING
```

Figure 7.4. Revised simulated annealing algorithm for macrocell thermal placement.

for calculating the thermal penalty:

$$\text{Penalty} = \alpha \cdot m \cdot [\sum_{i=1}^{m} \max((T[i] - T_{op})^3, 0) + \beta \cdot (T_{max} - T_{op})^3] \quad (7.18)$$

where m is the number of surface nodes, $T[\cdot]$ is the temperature profile of the current placement, T_{op} is the optimal surface temperature, T_{max} is the maximum surface temperature, and α and β are user controllable scaling factors.

The algorithm for macrocell thermal placement is given in Fig. 7.4. It is virtually the same as the placement algorithm in [9], except that the cost function now contains the extra thermal penalty term shown in Eq. (7.18) in addition to the original terms for area and wirelength. The algorithm can still handle designs with multiple power dissipation patterns; one only needs to calculate the temperature and insert an additional thermal penalty term for every power dissipation pattern individually.

7.6. SIMULATION EXAMPLES

The thermal placement tool based on the above algorithms can be used effectively. The programs were implemented in the C language. The cooling schedule for simulated annealing is adopted from [10].

The tool was applied to six standard cell (*biomed*, *primary1*, *primary2*, *sp1*, *struct* and *industry1*) and two macrocell benchmark circuits (*ami33* and *ami49*), and the simulation results on a machine with Intel Pentium 233MMX CPU and 64MB physical memory are listed in Tables 7.1 and 7.2. For the standard cell circuits, the cell power dissipation is estimated according to Eq. (7.17), with the rate of switching activities randomly generated between $0 \sim 1$ for each net. The input pin capacitance of each gate is assumed to be 0.1 pF, and the wire capacitance is assumed 242 pF/m. The clock frequency is assumed to be

Table 7.1. Standard cell thermal placement simulation results.

name	# cells	# nets	Original T_{max} (°C)	Original L_{total} (μm)	Original CPU (sec)
biomed	9519	7052	66.2	4025380	3413
primary1	2446	1266	43.7	827552	181
primary2	5103	3817	31.8	3749190	604
sp1	1268	915	29.8	1023116	514
struct	3070	1920	34.6	1115994	546
industry1	3012	2594	70.4	1387875	726

Table 7.1 (continued). Standard cell thermal placement simulation results.

name	Thermal Placement T_{max} (°C)	Thermal Placement L_{total}(%)	Thermal Placement CPU (%)
biomed	61.7	3884820 (93%)	5107 (149%)
primary1	40.1	785821 (95%)	261 (144%)
primary2	28.2	3646385 (97%)	827 (136%)
sp1	25.5	1054690 (103%)	709 (137%)
struct	29.3	1087371 (97%)	694 (127%)
industry1	63.9	1351899 (97%)	984 (135%)

Table 7.2. Macrocell thermal placement simulation results.

name	# cells	# nets	Original T_{max} (°C)	Original L_{total} (μm)	Original area (m^2)	Original CPU (sec)
ami33	33	131	46.8	143223	3.15e-6	13
ami49	49	408	79.73	1925470	6.24e-5	27

800 MHz. For the macrocell circuits, the cell power dissipation is randomly generated by assigning cells power densities ranging from 2.2×10^4 (W/m^2) to 2.4×10^6 (W/m^2), which are the typical values in modern high performance circuits such as microprocessors. Thermal conductivity of the package is assumed to be 7 (W/m°C) for the sides, 2000 (W/m°C) for the top, and 8800 (W/m°C) for the bottom for all cases.

The finite-difference mesh used for the substrate thermal modeling has 20 $\sim$ 40 grid lines in the X and Y directions, and 6 in the Z direction. The runtime of the pre-characterization phase (including substrate thermal model reduction,

Table 7.2 (continued). Macrocell thermal placement simulation results.

name	Thermal Placement T_{max} (°C)	Thermal Placement L_{total}(%)	Thermal Placement area (%)	Thermal Placement CPU (%)
ami33	45.5	154167 (107.6%)	3.18e-6 (101%)	31 (238%)
ami49	50.29	2028220 (105.3%)	6.54e-5 (104.8%)	72 (267%)

thermal/power objectives construction, etc) varies between 1 $\sim$ 5 minutes. In Tables 7.1 and 7.2, T_{max} is the maximum on-chip temperature, L_{total} is the estimated total wire length, and CPU is the execution time of the placement tool. The net length is estimated by using the half-perimeter bounding box model. The numbers in parentheses are the thermal placement results in percentage term of the traditional placement results, in which only the total wirelength and area are used as the optimization objectives.

Overall, the thermal placement algorithms provide noticeable improvements in thermal distribution in the final layouts. An execution time overheads of 30% $\sim$ 50% were observed for the case of standard cell thermal placement. This runtime overhead primarily stems from the increased cell movement attempts due to rejecting cell movements that will worsen the thermal constraint violations. For standard cell thermal placement, no area increase was observed for all circuits, and on average the thermal placement algorithm yields approximately the same or slightly smaller total wire lengths than using the traditional placement approach. One possible explanation of the slightly better wire length results is that rejecting thermal constraint violating cell movements effectively results in a slower annealing schedule, which tends to produce better results if the original annealing process is not absolutely optimal.

In view of wirelength and area, the results for macrocell thermal placement are showing more overhead in comparison with the case of standard cell placement. The runtime overheads range between 140% $\sim$ 170%, which come primarily from the matrix multiplications in estimating the incremental temperature profile change, and also from the evaluation of Eq. (7.18). Final areas increase only slightly ($< 5\%$), but the wirelength increases are between 5% $\sim$ 10% even after extensive tweaking of Eq. (7.18). Without tweaking, the wirelength and final area could easily increase up to 30%. It is obvious that the tradeoff does exist between the traditional and thermal objectives after the addition of the thermal penalty term in the simulated annealing cost function. Currently no reported implementation strategy that is based on power distribution can capture the thermal coupling effects, without wirelength or area degradation when taking the thermal profile quality into placement consideration. This is a subject worth further investigation.

To illustrate the thermal distribution improvement, the temperature profiles of *ami49* before and after thermal placement are shown in Fig. 7.5(a) and Fig. 7.5(b). The histograms of the temperature values at the surface grid points for all benchmarks are given in Fig. 7.6 - Fig. 7.13. Note that in all simulations the ambient temperature is assumed to be zero, therefore all the temperature values shown in the figures are caused by cell power dissipation alone.

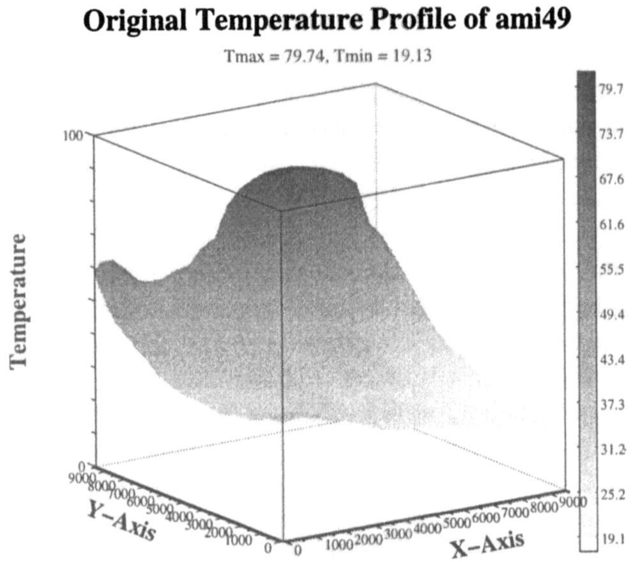

Figure 7.5(a). Temperature profiles of benchmark *ami49* without thermal placement. The ambient temperature is assumed to be zero.

7.7. SUMMARY

This chapter describes two cell placement tools for improving the substrate thermal distribution of both standard cell and macro cell designs.

- An brief overview of existing thermal placement studies are given.
  - It is pointed out that the uniform power distribution does not guarantee uniform temperature distribution.
  - A limiting factor of most existing thermal placement tools is the assumption of constant power dissipation, which is not true when the loading capacitance changes during placement.
- The substrate temperature calculation that was introduced in Chapter 4 is revisited in this chapter. The finite-difference equations for all nodes in the thermal system are represented in a matrix form. This matrix form serves as the basis for later compact substrate thermal modeling.
- The concept of the compact substrate thermal modeling is presented.
  - The compact substrate thermal model significantly improves the efficiency of the temperature profile estimation and optimization.

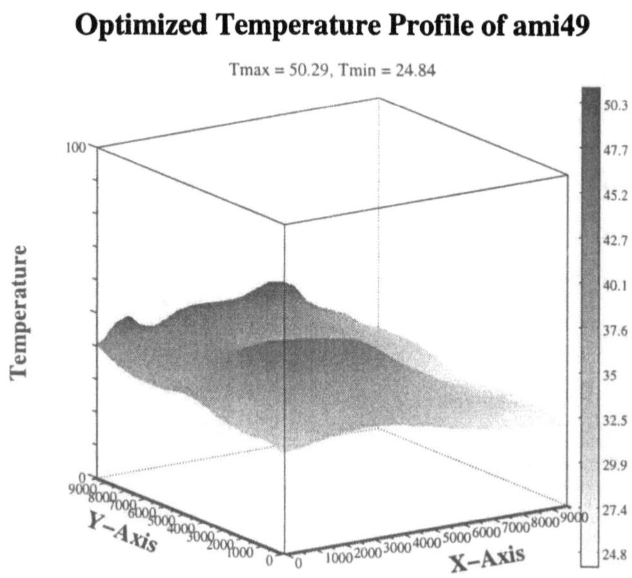

Figure 7.5(b). Temperature profiles of benchmark *ami49* with thermal placement. The ambient temperature is assumed to be zero.

- Two approaches for deriving the compact thermal models are described and their runtime efficiencies are compared. The first approach uses the superposition principle to construct the transfer thermal resistance matrix. The second approach directly manipulates the nodal admittance matrix.

- Two thermal placement algorithms that utilize the compact substrate thermal modeling are discussed; one is for the standard cell placement and the other is for the macrocell placement.

  - For the standard cell thermal placement, a new simulated annealing algorithm is developed and presented. During the annealing process, the power budget that constrains the cell movement is gradually tightened according to the cooling schedule.
  - For the macrocell thermal placement, it is pointed out that the thermal coupling effect between macrocells is dominant. A thermal penalty term is added to the cost function during placement to discourage uneven temperature distribution.

- The above thermal placement algorithms can be applied to the designs with multiple power dissipation patterns, such as sequential circuits or

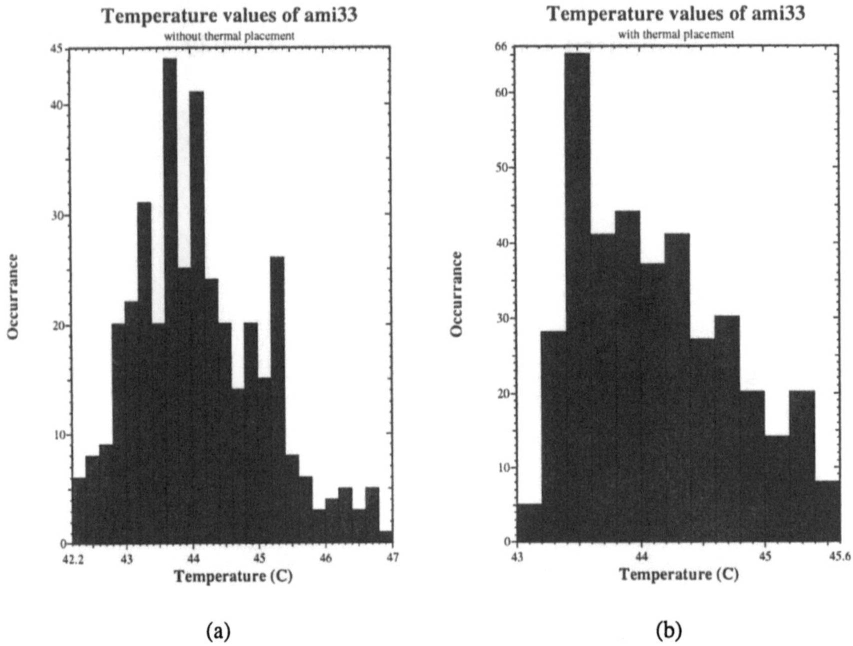

Figure 7.6. Histograms of on-chip temperatures of *ami33* (a) before and (b) after thermal placement.

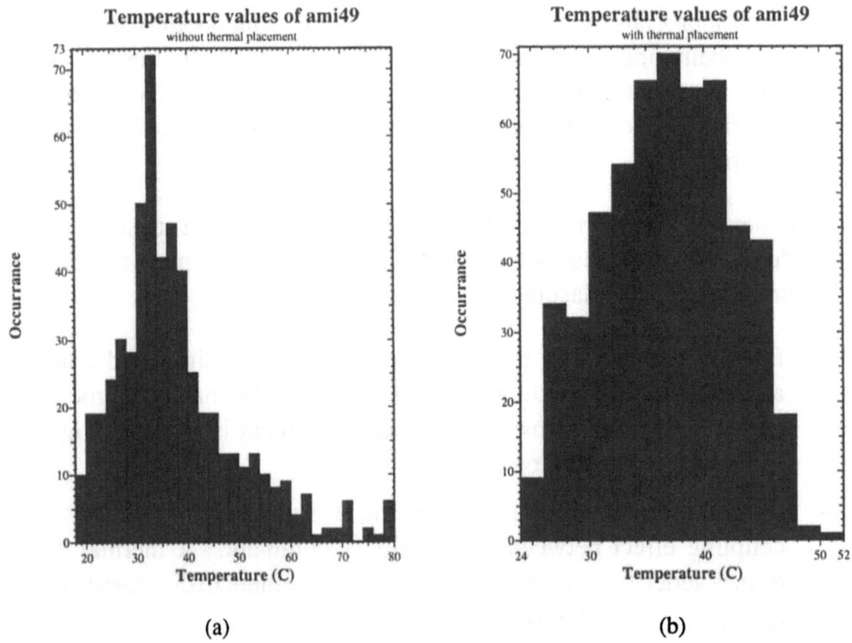

Figure 7.7. Histograms of on-chip temperatures of *ami49* (a) before and (b) after thermal placement.

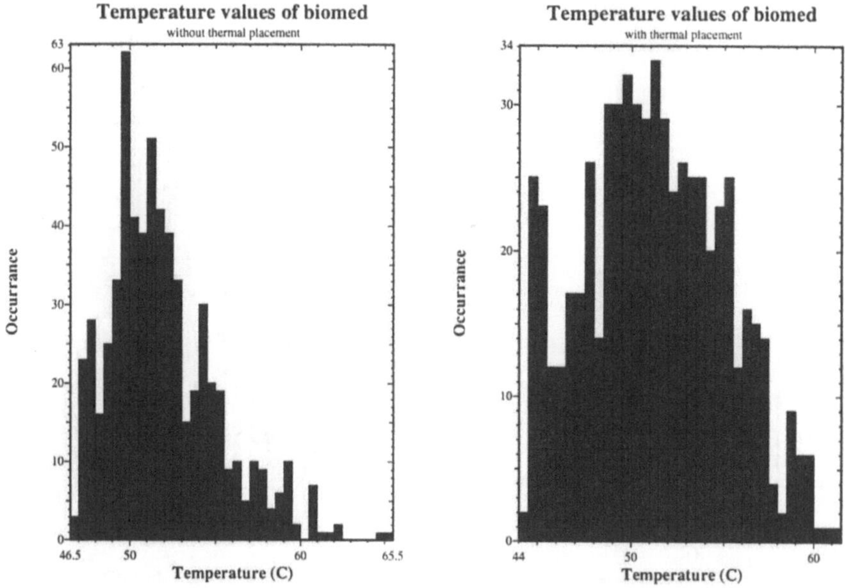

(a) (b)

Figure 7.8. Histograms of on-chip temperatures of *biomed* (a) before and (b) after thermal placement.

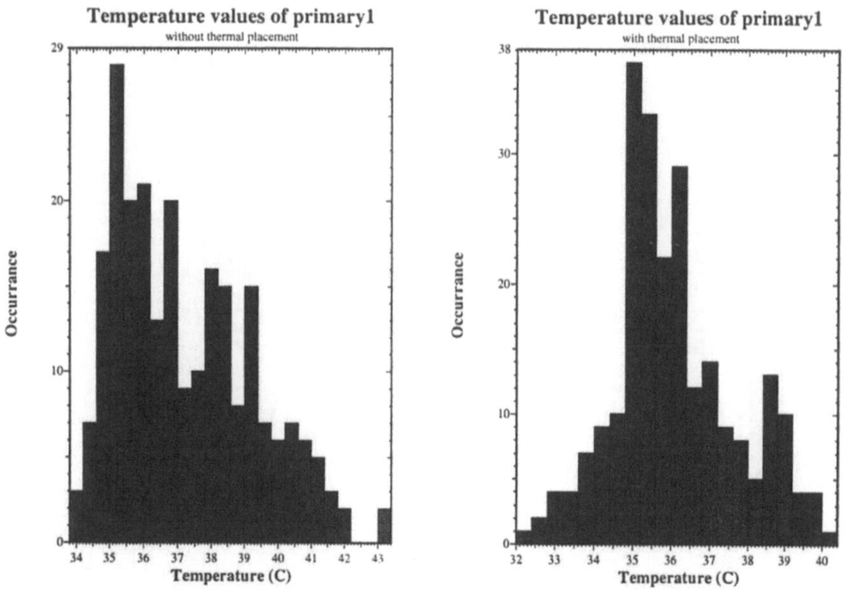

(a) (b)

Figure 7.9. Histograms of on-chip temperatures of *primary1* (a) before and (b) after thermal placement.

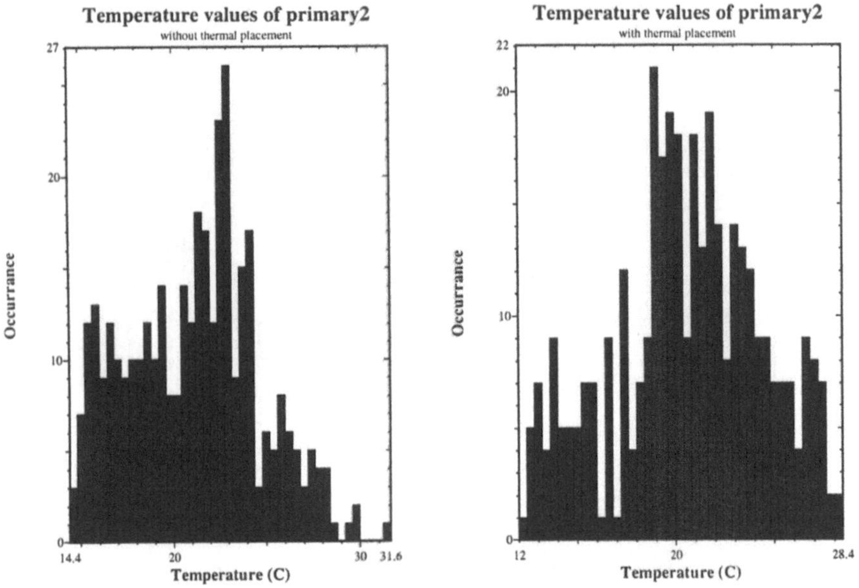

(a) (b)

Figure 7.10. Histograms of on-chip temperatures of *primary2* (a) before and (b) after thermal placement.

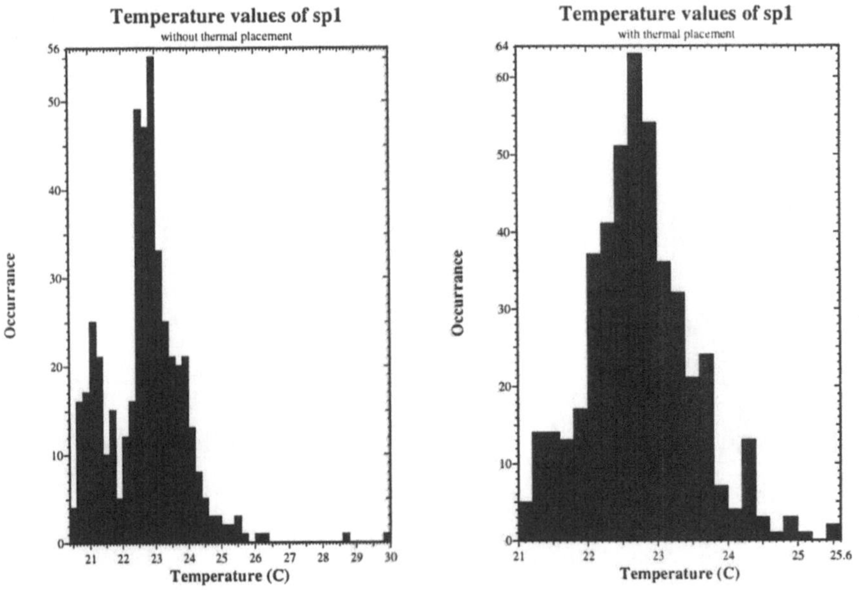

(a) (b)

Figure 7.11. Histograms of on-chip temperatures of *sp1* (a) before and (b) after thermal placement.

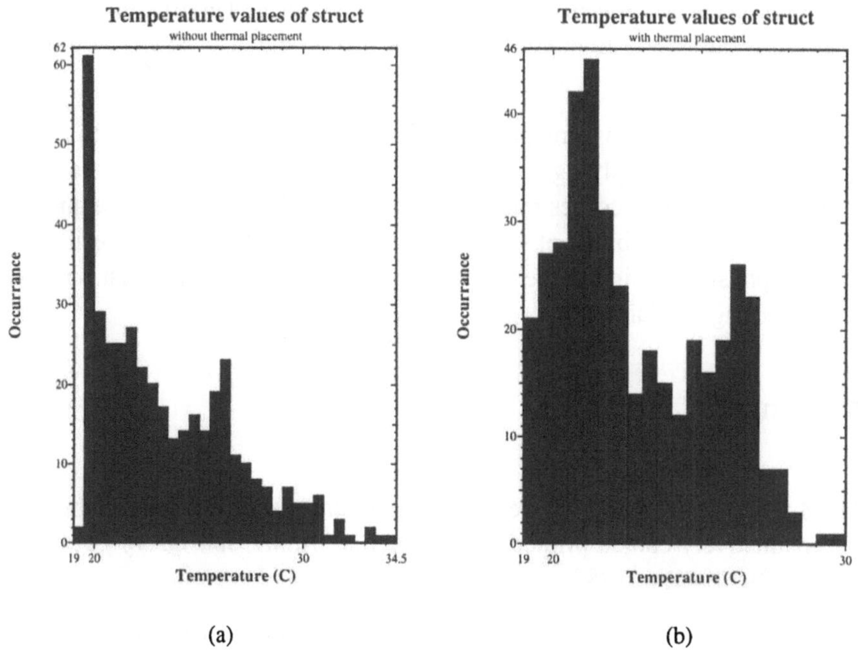

Figure 7.12. Histograms of on-chip temperatures of *struct* (a) before and (b) after thermal placement.

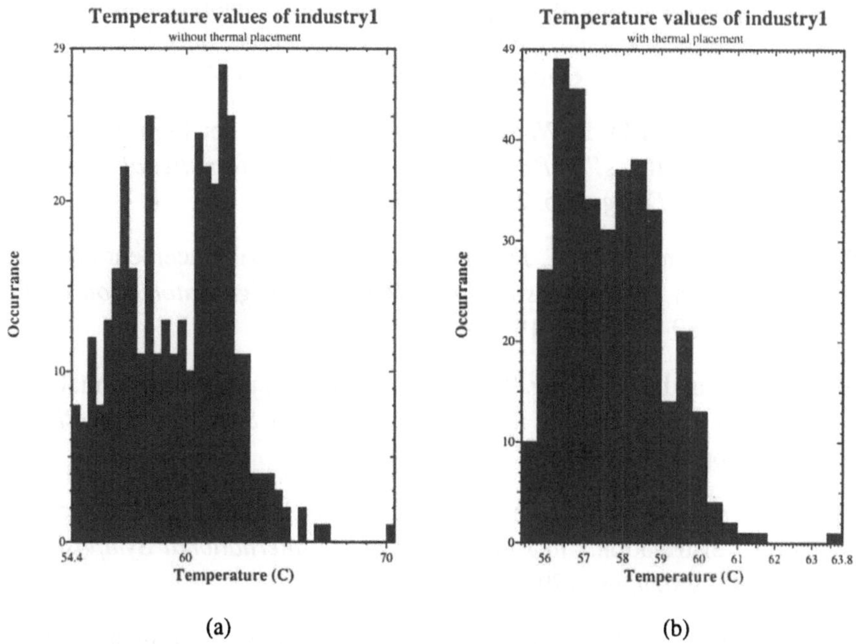

Figure 7.13. Histograms of on-chip temperatures of *industry1* (a) before and (b) after thermal placement.

microprocessors. It is useful in controlling the local operating temperatures at temperature sensitive subcircuits for mixed-signal or system-on-a-chip designs.

- Simulation examples are provided for both standard cell and macrocell placement. By applying the thermal placement algorithms, the temperature distribution becomes more uniform with little impact on area and wire length. The possibility of extending the thermal placement algorithms to other physical design processes such as floorplanning and netlist partitioning/routing is addressed.

References

[1] M. D. Osterman and M. Pecht, "Component placement for reliability on conductively cooled printed wiring boards," *ASME Journal of Packaging*, 111(3):149–156, 1989.

[2] R. Darveaux, I. Turlik, L. T. Hwang, and A. Reisman, "Thermal stress analysis of a multichip package design," *IEEE Transactions on Components, Hybrids, and Manufacturing Technology*, pp. 663–672, Dec. 1989.

[3] M. D. Osterman and M. Pecht, "Placement for reliability and routability of convectively cooled PWBs," *IEEE Transactions on Computer-Aided Design of Integrated Circuits and Systems* , 9(7):734–744, Jul. 1990.

[4] K. Y. Chao and D. F. Wong, "Thermal placement for high performance multi-chip modules," in *Proceedings of the IEEE International Conference on Computer Design*, pp. 218–223, Oct. 1995

[5] H. Eisenmann and F. M. Johannes, "Generic global placement and floorplanning," in *Proceedings of the ACM/IEEE Design Automation Conference*, pp. 269–274, June 1998.

[6] C. N. Chu and D. F. Wong, " A matrix synthesis approach to thermal placement," in *Proceedings of the 1997 International Symposium on Physical Design*, pp. 163–168, 1997.

[7] C. H. Tsai and S. M. Kang, "Standard cell placement for even on-chip thermal distribution," in *Proceedings of the International Symposium on Physical Design*, pp. 179-182, April 1999.

[8] C. H. Tsai and S. M. Kang, "Macrocell placement with temperature profile optimization," in *Proceedings of the International Symposium on Circuits and Systems*, pp. 390–393, 1999.

[9] C. Sechen and A. Sangiovanni-Vincentelli, "The TimberWolf placement and routing package," *IEEE Journal of Solid-State Circuits*, Vol. SC-20, No. 2, April 1985, pp. 510–522.

[10] M. Huang, F. Romeo and A. Sangiovani-Vincentelli, "An efficient general cooling schedule for simulated annealing," in *Proceedings of the International Conference on Computer-Aided Design*, pp. 381–384, Nov. 1986.

Chapter 8

TEMPERATURE-DRIVEN POWER AND TIMING ANALYSIS

8.1. INTRODUCTION

It has been shown in Chapter 3 that the MOS transistors are sensitive to their local temperatures. The carrier mobility and the driving capability of the source-drain current are reduced at higher temperatures. As a result, the circuit performance can be considerably degraded if the on-chip temperature is not well controlled. This was also evidenced by the experiments shown in Chapter 5. Indeed, the avoidance of hot spots is exactly the reason why the thermal placement concept introduced in Chapter 7 is important.

Given the fact that the circuit delay is strongly dependent on temperature, one may want to know how the on-chip temperature gradient affects the overall chip timing. To be more specific, one may ask, "Does a critical path become less critical, or a non-critical path become critical because of the on-chip temperature gradient?"

This chapter addresses the above question. To find the steady-state temperature distribution for temperature-dependent timing analysis, the *statistical* power and temperature estimation techniques are used. As described in Chapter 2, the statistical power analysis is an efficient and accurate way for the average power estimation. Moreover, it is more meaningful to handle the environmental variables such as temperature in a statistical manner when the system timing is concerned.

Figure 8.1 shows the relationship between power, temperature, and timing in a VLSI system. The temperature variation directly impacts the power consumption (i.e., short-circuit power and leakage power) and delay. On the other hand, different power distribution can generate very different temperature profile. In order to accommodate the degraded timing due to temperature rise, the clock frequency must be adjusted, which in turn changes the power con-

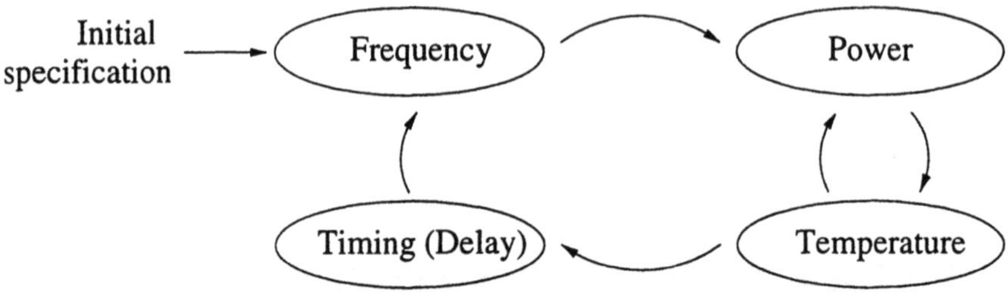

Figure 8.1. Relations between power, temperature, and timing.

sumption. Since the power, temperature, and timing are mutually related, this chapter discusses the temperature-driven power and timing analysis as a single subject.

In the following, the general overview of timing analysis techniques will be given first. The base methodology used in the statistical power and temperature estimation will be discussed next. Finally, the temperature-dependent timing analysis, including the delay modeling and the analysis results, will be presented.

8.2. TIMING ANALYSIS OVERVIEW

Timing analysis is one of the most critical tasks in high-performance ULSI system design. Over the last decade, designers have increasingly resorted to timing analysis tools to check whether a given circuit meets the performance goal (i.e., clock speed). Timing analysis of the ULSI design consists of checking for short and long path (*critical path*) problems.

Traditional timing analysis methods consist of two branches: *dynamic* and *static* methods. Not only are the underlying concepts of the two methods distinct, the delay models used can also be totally different. In the following, both methods will be discussed with the comparison of their pros and cons.

8.2.1 DYNAMIC TIMING ANALYSIS

Dynamic timing analysis is also called the *delay simulation*. It simulates a design with input patterns and collects the timing (delay) information. The simulation engines used in the dynamic timing analysis are similar to those used in the power analysis, which were discussed in Chapter 2.

Dynamic timing analysis approaches have been widely used for studying the timing of a design. After the simulation, the waveforms at primary outputs are inspected and the timing violations will be reported. The timing relationships such as *setup* and *hold* among the internal signals are also verified. Moreover, the dynamic timing analysis can be used for functional verification as well.

Normally in dynamic timing analysis, the input pattern that triggers the critical path can be easily identified with no extra computational cost. It is because the input patterns are exercised one at a time and the circuit timing is monitored for each pattern. Since it is always good to know which input pattern causes the timing to fail, the dynamic timing analysis can produce such information as its by-product. Another advantage of the dynamic timing analysis is that it avoids the false path problem. The definition of a false path will be given later when static timing analysis is discussed.

Because it is necessary to identify all possible timing errors in timing analysis, a complete set of input patterns need to be generated in this dynamic approach. Unfortunately, the generation of the complete set or the set that covers all possibilities can be difficult. Even if such a set is generated, it is impractical to simulate all input patterns in this set in an exhaustive way if the design is large or complex. Therefore, the dynamic timing analysis is primarily used for small circuits. It can also be used to accurately compute the delay of the paths that are known to be critical.

8.2.2 STATIC TIMING ANALYSIS

The static timing analysis identifies timing violations without the knowledge of input patterns, therefore it is much faster than the dynamic timing analysis. The *late mode* analysis (sometimes called "long path" analysis) propagates the latest arrival times for each logic block, which in turn finds the largest cumulative path delays. This mode identifies paths that will prevent the hardware from being able to operate at desired clock cycle time. The *early mode* analysis (sometimes called "short path" analysis) propagates the earliest arrival times for each block, which in turn finds the smallest cumulative path delays. This mode identifies paths that will cause the hardware to incorrectly store data into the previous clock cycle. In the rest of this chapter, the late mode analysis will be assumed for the convenience of discussion, unless otherwise indicated.

In general, a static timing tool requires the following information as its inputs:

- the points in the logic model where the timing is of interest (e.g., inputs, outputs, logic gates)
- the timing relationship between those points of interest (e.g., delays, setup, hold)
- arrival times asserted at the logic inputs
- required arrival times asserted at the logic outputs
- clock definition (if sequential logic is present)

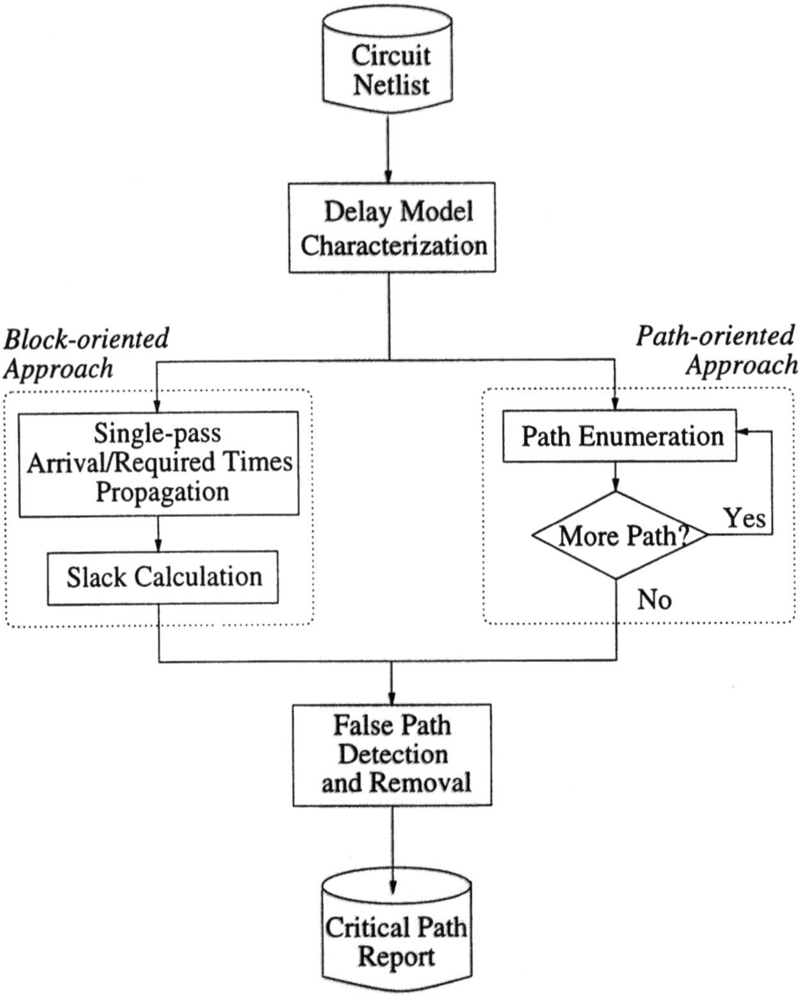

Figure 8.2. Block diagram of static timing analysis.

The block diagram of the general static timing analysis procedure is shown in Fig. 8.2. There are two different approaches in static timing analysis. The first one is the *path-oriented* approach, also called the path enumeration approach. The other one is the *block-oriented* approach. The path-oriented approach handles the timing problem for each unique path, while the block-oriented one does the same thing for each unique block (gate). In the following, these two approaches will be described. Because the block-oriented approach is faster and requires less memory, it is ideal for the VLSI systems, and will be the focus of the discussion.

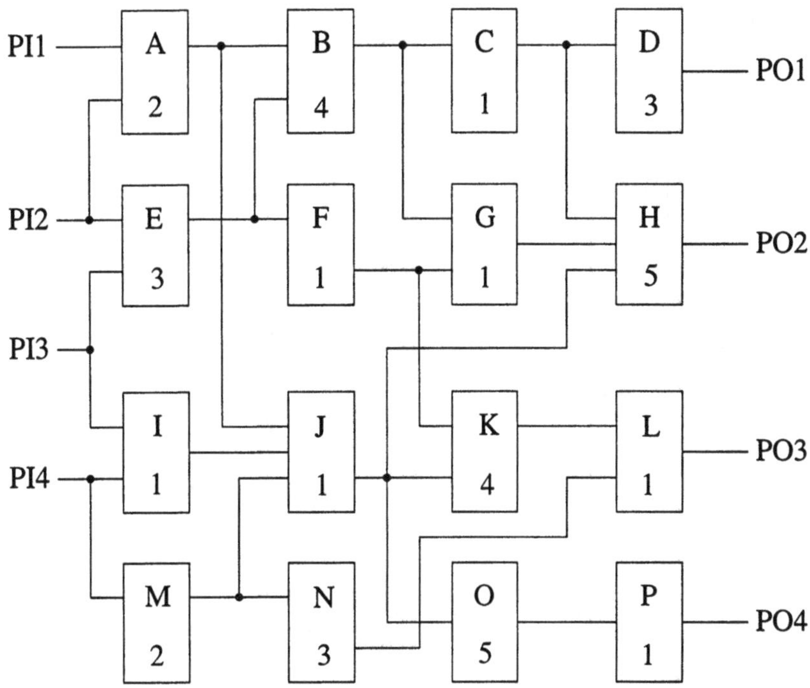

Figure 8.3. An example circuit diagram.

PATH-ORIENTED APPROACH

The path-oriented approach has been employed in many static timing analysis tools [1, 2, 3]. Consider the circuit diagram in Fig. 8.3 borrowed from [4]. The block delays are shown at the bottom of the blocks marked from A to P. The rising and falling delays are assumed to be identical for simplicity. To perform the path-oriented timing analysis, the most straightforward way is to enumerate all paths in the circuit from the primary inputs (PI1 $\sim$ PI4) to the primary outputs (PO1 $\sim$ PO4). For this small circuit, there are a total of 32 paths. Next, the path delays are found by adding up the delays of individual blocks. For instance, the delay of the path PI1-A-B-C-H-PO2 is 12.

The above approach that enumerates all paths is clearly expensive. It is impractical for circuits with larger size or more complex structure. An alternative is to extract the *k-most* critical paths. One example is to find only the most critical path [5], in which the depth-first search with pruning was used. The above algorithm is efficient, but extracting only one critical path often fails to provide enough information for correcting the timing violations. In 1989, Yen et al. developed an algorithm which traces the k-most critical paths [6] and the

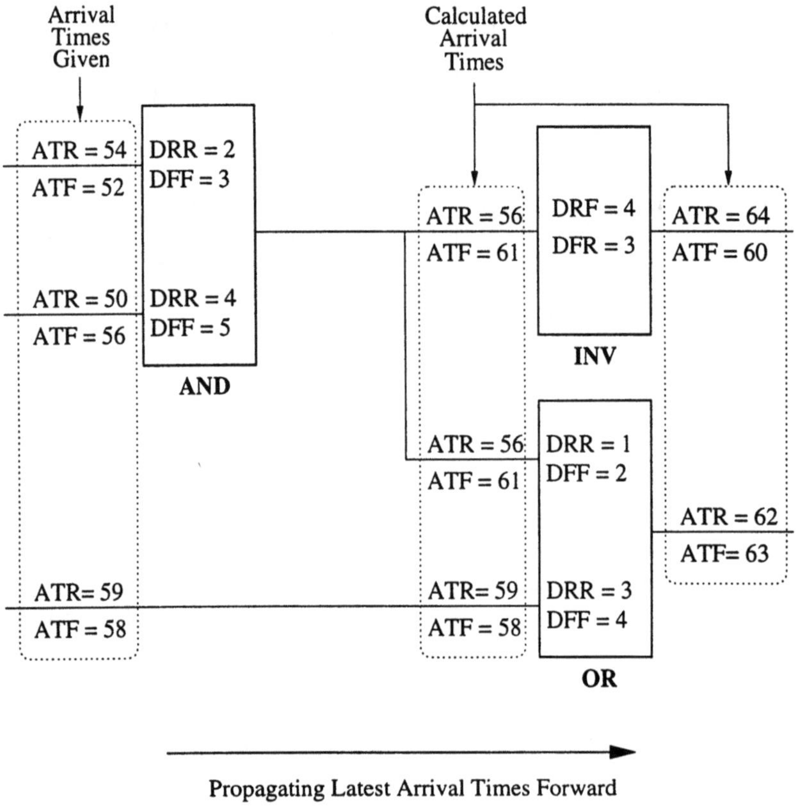

Figure 8.4. Arrival time propagation in block-oriented analysis.

sorted path delays were reported. A more efficient algorithm using the idea of branch slacks was later proposed to extract the k-most critical paths [7].

BLOCK-ORIENTED APPROACH

The above path-oriented approach finds timing problems one path at a time. For some logic blocks, like block K in Fig 8.3, there are many different paths passing through. In other words, these blocks are analyzed several times by the path-oriented approach.

A more efficient approach is to analyze all paths simultaneously. The blocks are analyzed in order and the *worst* timing caused by the paths passing through the block is recorded. It is therefore called the block-oriented approach. Consider a simple circuit in Fig 8.4, where three blocks (AND gate, OR gate, inverter) are shown. In Fig 8.4, ATR and ATF denote the rise and fall arrival times; DRR and DFF denote the the rise and fall block delays of the non-inverting logic; DRF and DFR denote that rise-to-fall and fall-to-rise block delays of the inverting logic.

In block-oriented timing analysis, the arrival times at the primary inputs are asserted as given. The arrival times of the rest of the circuit are then concurrently propagated forward to the primary outputs. During the propagation, the *latest* arrival times are recorded. For instance, the rise arrival times at the output of the AND gate is computed as $\max(54 + 2, 50 + 4) = 56$. Similarly, the fall arrival times is $\max(52 + 3, 56 + 5) = 61$.

In order to see whether the arrival times at the primary outputs are late, the *required* arrival times must be specified (asserted) at the primary outputs. If the arrival time is greater than the required arrival time at some primary output, there is at least one timing problem. Next, the required arrival times of internal nets are calculated by propagating the asserted required times backward to the primary inputs. During the propagation, the worst case must be considered and thus the *earliest* required arrival times are chosen for the late-mode analysis. Figure 8.5 illustrates the backward propagation, where RATR and RATF denote the rise and fall required arrival times.

Finally, to find out exactly which blocks are causing the timing problems, the concept of *slack* is used for convenience. The late-mode slack is defined as

$$slack = (\text{Required arrival time} - \text{Arrival time}). \tag{8.1}$$

If a net has negative slack, the signal is late. A slack calculation example for the sample circuit is given in Fig. 8.6, where SLKR and SLKF are the rise and fall slacks. Note that the slack value is constant $(= -3)$ along the worst path through the logic.

The block-oriented approach is fast. Moreover, the critical gates can be easily identified. It is well suited for integration with the logic synthesis program that requires values of the gate slack. However, this approach produces less information about the design timing than the path-oriented approach. The block-oriented analysis only records the worst slack for a given point of the logic. Therefore, unlike the path-oriented approach, it is difficult to handle the problem of finding the k-most critical paths.

DETECTION AND REMOVAL OF FALSE PATHS

The static timing analysis approaches need no knowledge of the input pattern and the functionality of the blocks (i.e., only need to know whether the blocks are inverting or non-inverting). Although computationally very efficient, they often lead to serious overestimation of the critical path delay due to the *false path* problem. A false path is not a true path along which signals can actually propagate. One example of a false path problem is shown in Fig. 8.7 (from [8]). Path $P = < b, d, e, x, y >$ is considered a false path because in order for

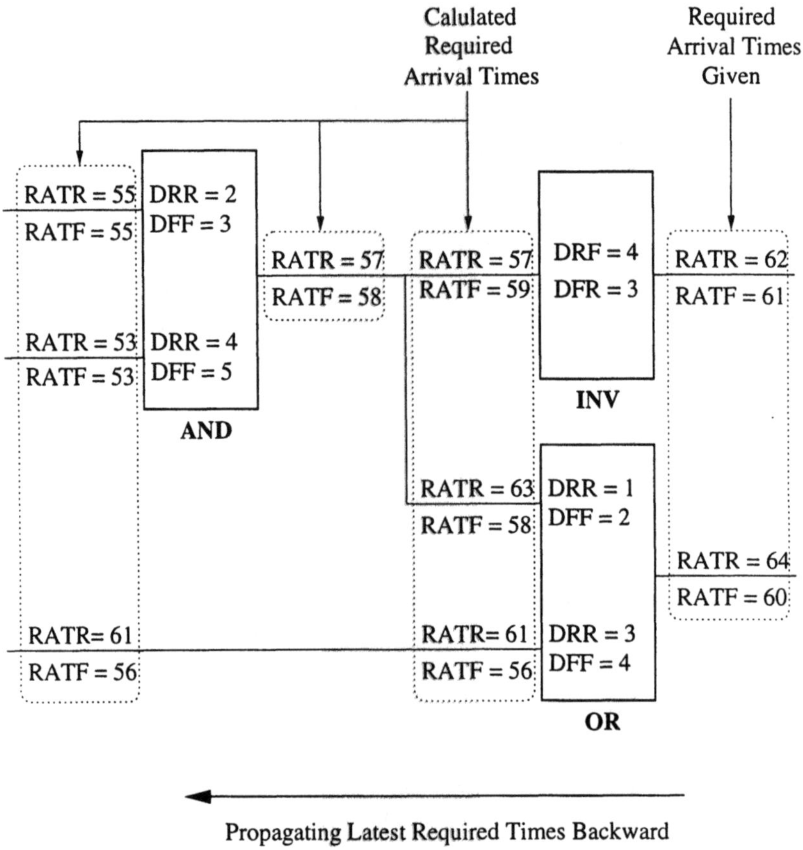

Figure 8.5. Required arrival time propagation in block-oriented analysis.

signals to propagate from gate *d* through gate *e*, *c* has to be 1 which blocks signals from x through gate y.

Several approaches have been developed to resolve the false path problem, and the most primitive one is called the *static sensitization*. A statically sensitizable path is the one that can be activated in isolation from other paths, with all of its side-inputs held at constant noncontrolling values (e.g., 1 for AND gates and 0 for OR gates). In [9], efficient algorithms and a backtracking technique have been utilized to find the statically sensitizable paths. A new idea that totally eliminates the backtracking process, which is usually very costly, has been proposed by Ju et al. [7]. It transforms the sensitization problem into a satisfiability problem and applies the binary decision diagram (BDD) [10, 11] to construct the output functions of the paths. Although the use of BDD package is often limited to small circuits, the satisfying sets for all of the internal nodes of the slowest primary output function can be constructed in a very short

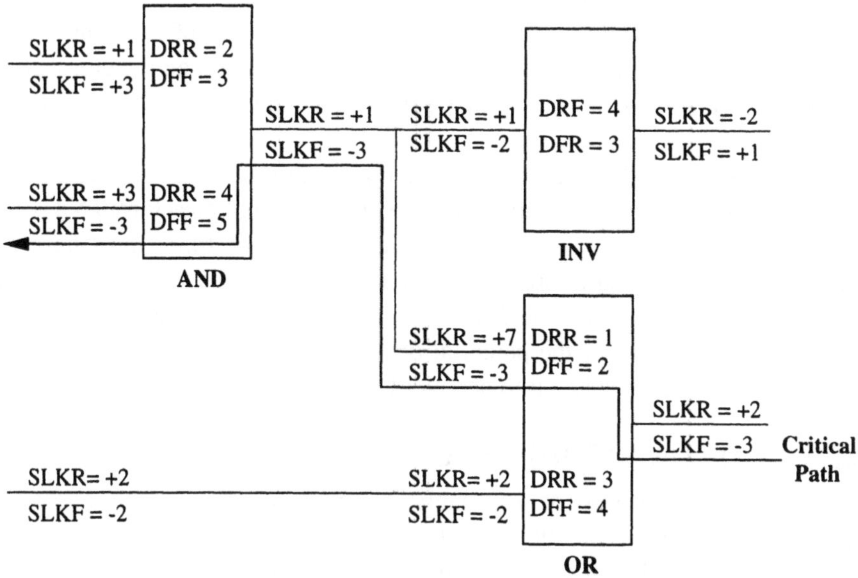

Figure 8.6. Slack calculation in block-oriented analysis.

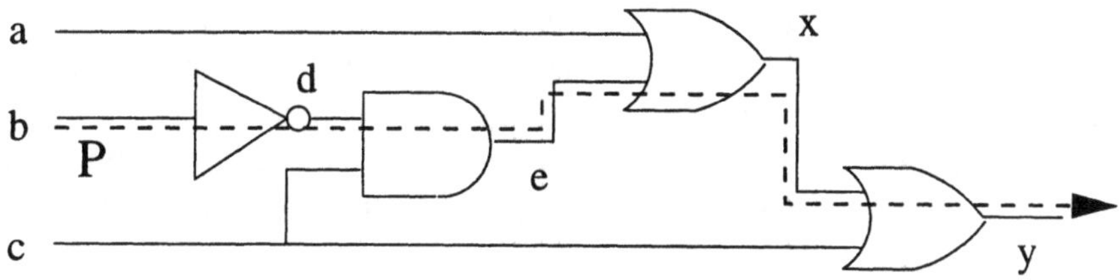

Figure 8.7. A false path example.

time. Other approaches for solving the false path problem, such as those based on the dynamic sensitization [12], the viability condition [13], and the Du's criterion [8], have also been proposed.

STATIC TIMING ANALYSIS FOR SEQUENTIAL CIRCUITS

The preceding examples and discussion focus on the static timing analysis of combinational logic. The theory directly applies to the analysis of synchronous sequential logic by breaking it into several combinational logics. The storage elements (latch, flip-flop) are usually chosen to be the break points. Because now the starting and/or ending points of the combinational paths are the storage elements, the timing assertions and constraints will come from the clocks that control the storage elements.

For systems with multiple clocks, the problem becomes much more complex. The timing analysis program needs to have a smart way of knowing which clocks launch and capture which data, and when to launch and capture them. The data may contain a clock *phase* tag besides their delay values for that purpose.

One other factor that makes the sequential timing analysis difficult is the *loop*. A loop exists in a sequential circuit when the signal goes through a series of transparent latches and feeds back to itself. A static timer must have the capability of automatically breaking such a loop, otherwise the propagation of arrival times and required arrival times will create infinite loops. A loop must be broken without hiding the potential timing problem, and the *slack stealing* effect also needs to be taken into account [14, 15, 16].

8.2.3 DELAY MODELING

The circuit delay must be accurately modeled in both dynamic and static timing analysis. For dynamic analysis, the delay accuracy is determined primarily by the simulation engine used. Chapter 2 and Chapter 5 have briefly introduced different kinds of simulation engines.

For static analysis, the base simulation unit is a block (gate). Therefore, the delay model of the gate must be carefully characterized before timing analysis is performed. The gate delay modeling with a single switching input has been addressed by many research works [17, 18, 19]. The case of multiple-input switching has also received much attention [20, 21, 22]. One general and powerful approach to model the gate delay is to numerically fit the SPICE-generated delay data by an empirical formula. This empirical formula is a function of the input slew, output loading, and a set of fitting parameters. For instance,

$$Delay = K_0 + K_1 \times C_{load} + K_2 \times T_{input\_slew} + \cdots. \tag{8.2}$$

This model is accurate, yet the lengthy and repetitive SPICE simulation is avoided in static timing analysis.

The gate delay is not only a function of the input slew and output loading, it is also affected by the temperature and voltage fluctuation, as well as the silicon process variation. The silicon process variation is difficult to capture and is often statistically modeled. The voltage fluctuation can be estimated by the IR-drop analysis tools. Traditionally the temperature effect is taken into account by assuming that a worst possible temperature value is *uniformly* distributed across the chip. It is a pessimistic assumption. It not only overestimates the gate delay and thus constrains the design space, but may also lead to timing problems by ignoring the on-chip temperature gradient. This issue will be addressed further later in this chapter.

8.3. STATISTICAL POWER DENSITY ESTIMATION

As described in Chapter 2, the power estimation method can be either input pattern dependent or independent. The pattern-dependent method is used when the input sequence is known for certain applications, and it produces a deterministic power value. However, input patterns are often unknown during the design phase. It is also impractical to estimate the average power by exercising all possible input patterns. As a result, in order to estimate the *nominal* on-chip steady-state temperature profile, it is more meaningful to calculate the average power in a statistical manner. This nominal temperature profile will later be used for the temperature-driven timing analysis.

A brief overview of the underlying theory of the statistical (Monte-Carlo) power analysis methods has been presented in Chapter 2. Interested readers may refer to it for more detail. In this chapter, a unique technique for the Monte-Carlo average power estimation, called the Mean Estimator of Density (MED) [23], is employed. The MED technique is a good mix of accuracy, speed, and ease of implementation. More importantly, it captures the transition statistics of each logic gate instead of the whole circuit. Therefore, it directly suits the purpose of temperature profile calculation.

Suppose a circuit is simulated n times, and for each time the number of logic transitions of a gate is x_i. According to the *central limit theorem* [24], the average $\overline{x} = \sum_{i=1}^{n} x_i/n$ has a distribution which is close to normal for large n. If μ is the true expected number of transitions of this gate, with $(1-\alpha)\times 100\%$ confidence it follows that

$$-z_{1-\alpha/2} \leq \frac{\overline{x}-\mu}{\sigma/\sqrt{n}} \leq z_{1-\alpha/2} \quad (8.3)$$

where σ is the standard deviation, and $z_{1-\alpha/2}$ is defined so that the area to its right under the standard normal distribution curve is equal to $\alpha/2$. Here we define $\overline{x}$ as the sample mean of the *power density* of a given gate in the circuit. Power density is defined as the power value per unit area, which is a direct measure of the local temperature rise. For sufficiently large number of n (i.e., $n \geq 30$), σ can be approximated by the sample standard deviation s. By using Eq. (8.3), one can show that the number of samples required is

$$n \geq \left(\frac{z_{1-\alpha/2}s}{\overline{x}\epsilon_1}\right)^2 \quad (8.4)$$

such that we have $(1-\alpha)\times 100\%$ confidence that the following condition is satisfied:

$$\frac{|\overline{x}-\mu|}{\mu} \leq \frac{\epsilon_1}{1-\epsilon_1} = \epsilon \quad (8.5)$$

where ϵ is the user-specified error tolerance. Equation (8.4) provides a stopping criterion to yield the power estimation accuracy specified in Eq. (8.5) with confidence $(1-\alpha)\times 100\%$.

It is clear from Eq. (8.4) that for a small value of $\overline{x}$, the number of samples required can be very large to meet the specified accuracy level. In MED-like approaches, the stopping criterion Eq. (8.4) is used for gates that have $\overline{x}$ larger than the user-specified threshold value, μ_{min}. These gates are referred to as the *regular-density* gates. A different stopping criterion is used for gates that have $\overline{x}$ less than μ_{min}:

$$n \geq (\frac{z_{1-\alpha/2^S}}{\mu_{min}\epsilon_1})^2. \tag{8.6}$$

These gates are referred to as the *low-density* gates. Equation (8.6) controls the number of samples by providing an absolute error bound for the low-density gates. Although the estimated power values of those gates are less accurate, they have the least effect on temperature rise.

8.4. MONTE-CARLO POWER-TEMPERATURE ITERATION SCHEME

To determine the nominal steady-state temperature profile of the chip substrate, the power values of the gates obtained from the above statistical simulation need to be input to a 3-D thermal simulator. Since power and temperature are functions of each other as described at the beginning of this chapter, an iteration scheme is invoked between the power density and temperature calculations [25].

The iteration scheme is graphically shown in Fig. 8.8. There are two levels of iteration. The first level is related to the Monte-Carlo power estimation, and the second level is related to the mutual dependence between power and temperature. In Fig. 8.8, the convergence rates of the first and second levels of iteration are determined by the quantities circled by the dashed lines. The quantities in <·> are non-constants, which are calculated and updated at run time.

The stopping (convergence) criterion of the first level of iteration is described by Eq. (8.4) and Eq. (8.6). The stopping criterion of the second level of iteration is based on two factors: the temperature difference between two consecutive iterations and the power estimation error inherited from the Monte-Carlo simulation.

Suppose the confidence level $(1-\alpha) \times 100\%$ and the percentage error ϵ are used in the Monte-Carlo power estimation for the regular-density gates. After the power estimation is complete in (second-level) iteration k, the power of each regular-density gate is compared with the one calculated in iteration $k-1$. The number of regular-density gates that have percentage power difference less than $2\epsilon/(1-\epsilon)$ between iterations $k-1$ and k is counted and denoted n_{rs}. If the ratio of n_{rs} to the total number of regular-density gates in the circuit is larger than $(1-2\alpha)$, the iteration process is stopped. Otherwise the thermal simulation is performed based on the power distribution in iteration k, and the

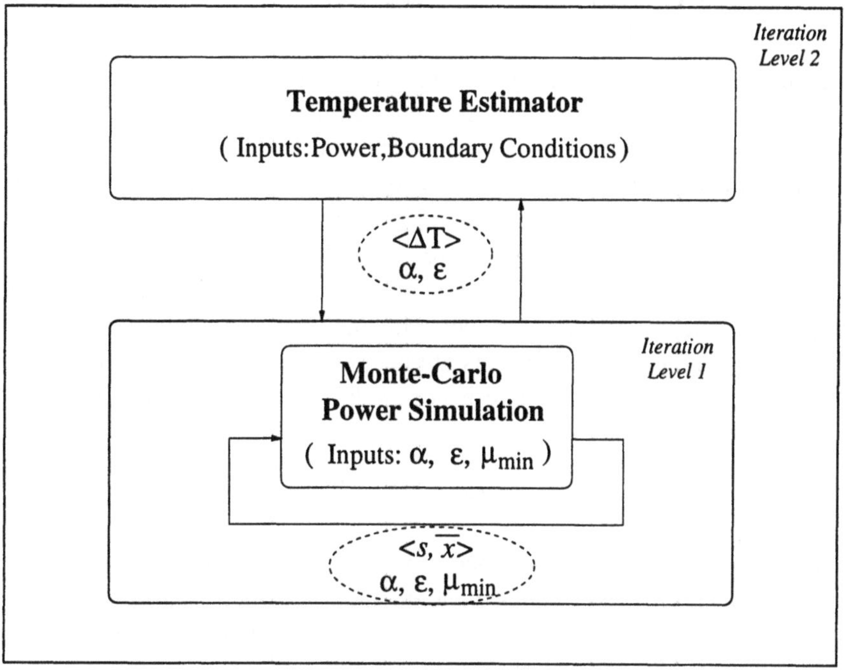

Figure 8.8. Monte-Carlo power and temperature iteration scheme.

updated temperature profile is found. The resulting temperature of each gate is then compared with the one in iteration $k-1$ in order to determine whether or not the iteration process can be stopped according to the user-specified accuracy level of temperature. The above $2\epsilon/(1-\epsilon)$ term accounts for the possible over-estimation and under-estimation of the power values inherited from using the Monte-Carlo approach. The value $(1-2\alpha)$ places an upper bound for the temperature effect to be considered important during iterations.

The above two-level iteration scheme was adopted in [25]. In this work, the external spatial correlation of the input signal vector is not considered in Monte-Carlo power estimation. The circuit is given a sequence of two input vectors for one logic simulation run. All possible input patterns (high, low, high-to-low, low-to-high) are assumed to have an equal probability of occurrence. Moreover, the logic simulator used in [25] takes as inputs the load capacitances, the input signal slope, and the temperature-dependent MOS device and interconnect parameters of each gate, as will be described in the following section. The state equations of the gates are formulated as the Riccati differential equations and solved analytically [26]. The above process is fast enough to make the temperature-sensitive statistical power estimation both accurate and feasible.

8.5. TEMPERATURE-DEPENDENT GATE AND RC DELAYS

From the experimental results presented in Chapter 5, it can be seen that the on-chip temperature gradient and temperature rise substantially affect the circuit delay. The critical path timing, affected by the delays of logic gates and interconnects, is therefore strongly temperature-dependent. Temperature-sensitive timing analysis is important to the high-performance VLSI design, and the assumption of uniform temperature across the chip may not be appropriate.

In [25], the temperature-dependent gate delay is calculated by using the regionwise quadratic (RWQ) model with the mobility model introduced in Chapter 3. As for the interconnects at given temperatures, the following equation is used in order to find the resistance value for temperature-dependent RC delay estimation:

$$R(T) = R_0[1 + \alpha_T(T - T_0)]. \tag{8.7}$$

In Eq. (8.7), $R(T)$ is the resistance at temperature T, R_0 is the resistance at room temperature T_0, and α_T is the temperature coefficient of resistivity (e.g., $0.004\,^\circ\mathrm{C}^{-1}$ for Aluminum). As a general rule of thumb, the RC delay increases about 5% for 10 °C of interconnect temperature rise.

To find the signal-line interconnect temperature, the coordinates of each metal are first extracted. Next, the localized substrate temperature at (x, y) is used as the temperature of the interconnects located near (x, y). It implies that the temperature difference between the substrate and the multi-layered signal-line metals is ignored. Note that, however, the Joule heating effect may also need to be taken into account separately for calculating the temperature of the multi-layered interconnects. Details of the interconnect temperature calculation considering Joule heating was described in Chapter 6.

To facilitate the RC delay calculation, the layout extractor developed in [25] extracts the signal-line interconnect resistance in the form of a distributed RC tree, as shown in Fig 8.9. Each signal-line interconnect tree is transformed into an equivalent π-model and is lumped to the corresponding driving gate. This is shown in Fig. 8.10, where T_g is the gate temperature, and $R_i(T_i)$ is the temperature-dependent resistance calculated by using Eq. (8.7).

8.6. SIMULATION EXAMPLES

Finding the path timing requires either dynamic or static timing analysis. In this section, the dynamic analysis will be used first to investigate how temperature can affect the timing and change the criticality of a path.

To dynamically find the critical (longest) path, it is assumed that the pool of all possible input patterns is provided. Therefore, the input pattern that triggers the critical path, i.e., critical pattern, must also be in this pool.

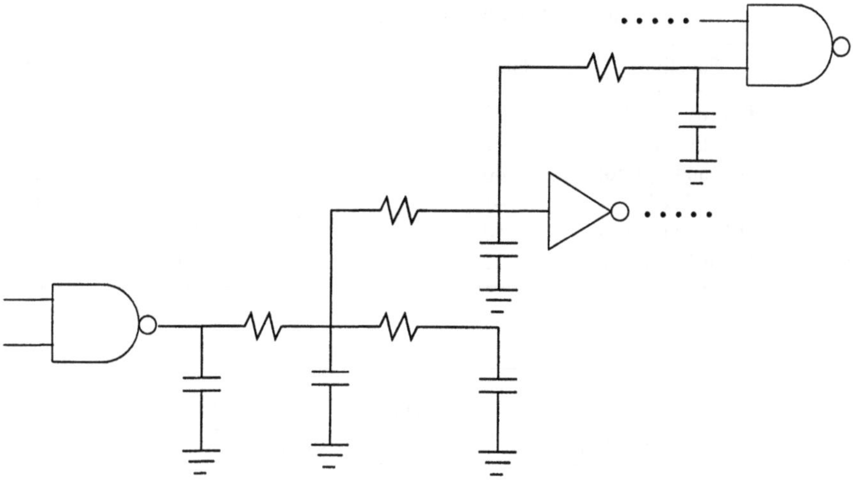

Figure 8.9. Example of a distributed RC tree.

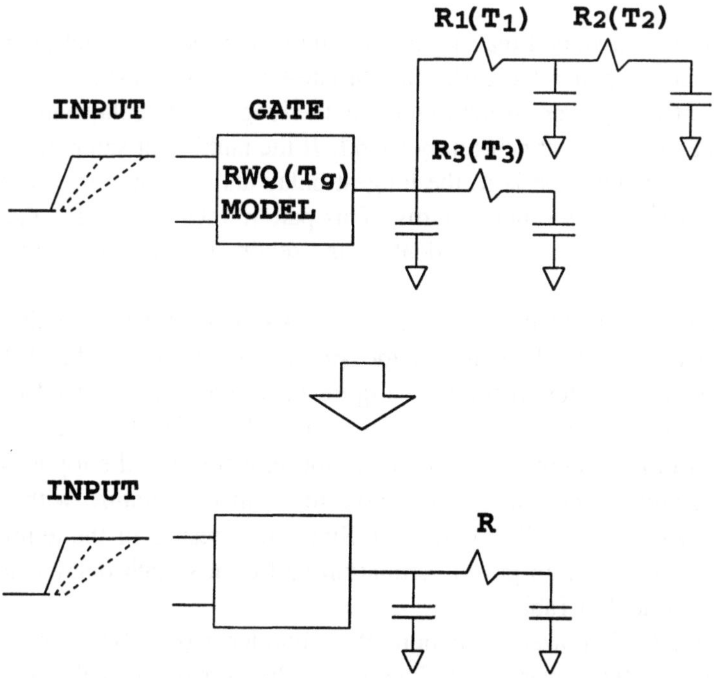

Figure 8.10. Example of an equivalent π model.

Table 8.1. The ISCAS85 benchmark circuits.

Circuit	Function	#inputs	#outputs	#transistors	#gates
C432	Priority decoder	36	7	1152	364
C499	ECAT	41	32	2266	719
C880	ALU and Control	60	26	1768	538
C1355	ECAT	41	32	2442	703
C3540	ALU and Control	50	22	5842	1656
C6288	16-bit Multiplier	32	32	10706	3001

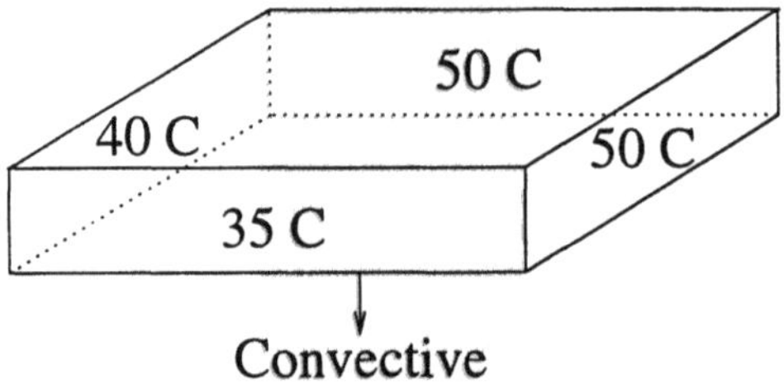

Figure 8.11. Thermal boundary conditions for temperature-dependent timing simulation.

In [25], it is assumed that the input-pattern pool is composed of the input patterns that are generated earlier for Monte-Carlo power estimation. During the Monte-Carlo power estimation phase, the longest path delay and its associated input pattern are concurrently found. If the number of samples needed in Monte-Carlo simulation is n, the longest delay and its triggering pattern will be found out of the n input patterns. This pattern thus obtained is the critical pattern, which will be used to identify and report the gates along the critical path.

In the remainder of this section, six ISCAS85 benchmark circuits will be used as examples to demonstrate the simulation results. Table 8.1 shows these circuits and their functions. Figure 8.11 shows the thermal boundary conditions used for all circuits under simulation [25]: The four sides are set to be in the isothermal condition, i.e., constant temperatures, the top is perfectly insulated, and the bottom is convective to room temperature with the heat transfer coefficient $5,000$ (W/m^2 °C). Simulation results of the temperature-dependent Monte-Carlo power estimation and critical path delay calculation are demonstrated in Table 8.2.

In Monte-Carlo power simulation, 95% confidence ($\alpha = 0.05$) and 5% error tolerance ($\epsilon = 0.05$) were used. The μ_{min} value was dynamically determined in the following way. Circuits were first simulated using an initial large μ_{min}. This provided a rough estimation of the power density distribution of the gates.

Table 8.2. Simulation results with dynamic timing analysis.

Circuit	Power	T_{dccb_max}	T_{dccb_min}	Delay(27 °C)
Unit	mW	°C	°C	ns
C432	7.1	47.69	36.15	5.66
C499	15.0	49.90	37.14	4.59
C880	11.1	49.43	38.29	3.92
C1355	12.8	51.02	36.20	4.52
C3540	41.1	49.88	35.38	6.87
C6288	380.7	50.04	35.23	14.71

Table 8.2 (continued). Simulation results with dynamic timing analysis.

Circuit	Delay(T)	Path Change?	MC CPU	Therm CPU
Unit	ns	-	sec	sec
C432	6.68		228.4	63.8
C499	5.51	√	397.6	131.6
C880	4.67		340.7	96.3
C1355	5.52		656.8	413.9
C3540	8.39	√	1522.9	1331.2
C6288	17.8	√	10094	3494

The new μ_{min} value was then chosen such that 10% of the gates were classified as low-density gates (i.e., with power density less than μ_{min}). The simulation was then rerun based on the new μ_{min} value. The estimated circuit powers are shown in the second column in Table 8.2. Here, T_{dccb_max} and T_{dccb_min} are the simulated maximum and minimum temperatures of the gates on the longest path, respectively. The longest path delays without considering the temperature effect are shown in the fifth column. The estimated temperature-dependent longest path delays (i.e., Delay(T)) are listed in the sixth column for comparison.

For a given critical input pattern, the critical path may be different for a circuit subject to a uniform room temperature and subject to a non-uniform temperature distribution. The circuits under simulation with changing critical path due to the thermal effect are marked with $\surd$ in the seventh column. The CPU times (on SUN SPARCstation 10) used for the Monte-Carlo power and thermal simulations are given in the last two columns of Table 8.2. Finally, the temperature profile of C6288 is demonstrated in Fig. 8.12. The gates on the longest path of C6288 are shown as small diamonds.

The static timing analysis is also performed on the same circuits and the results are given in Table. 8.3. In the static timing analysis, each gate has four

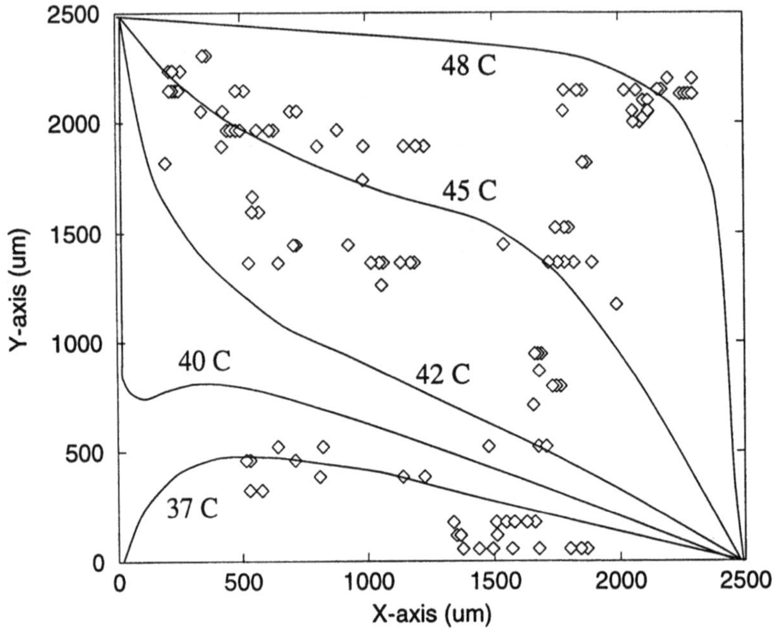

Figure 8.12. The simulated temperature profile and the gate distribution of the longest path in C6288: The solid lines are the isothermal temperature contour and the small diamonds are the on-chip locations of gates in the longest path.

Table 8.3. Simulation results with static timing analysis.

Circuit	Delay(27 °C)	Delay(T)	Path Change?
Unit	ns	ns	-
C432	8.49	9.58	
C499	6.75	8.12	√
C880	6.92	8.24	
C1355	7.47	9.33	√
C3540	10.99	14.43	√
C6288	25.31	30.25	√

delay values: t_rise(27 °C), t_fall(27 °C), t_rise(T_g), t_fall(T_g), where T_g is the gate temperature obtained directly from previous thermal simulation. It is assumed that each gate is subject to constant input slope and output loading. The four delay values of each gate are precharacterized and tabulated before timing analysis starts. When a gate is precharacterized, its loading interconnects are lumped to the output node of this gate and their temperatures-dependent resistances are used (See also Fig. 8.10).

The delay values in Table 8.3 are different from those in Table 8.2. The difference comes from two sources. Firstly, the delay models used in the dynamic and static timing analysis are different. Secondly, the input patterns used in the dynamic timing analysis are not complete, therefore the critical path found may not be the true critical path. The static timing analysis used for this example eliminates the false paths by using the backtracking technique [9].

Note that the temperature-induced critical path change also occurs in static timing analysis. In Table 8.3, one more circuit (C1355) changes its critical path because of the on-chip temperature gradient. The simulation results again confirm that the path delay must be accurately determined based on its local temperature, and the traditional assumption of the uniform temperature distribution may lead to false prediction of the timing problems.

8.7. SUMMARY

This chapter discusses the importance of the thermal effect on circuit timing. Because temperature distribution is determined by power distribution, a statistical power analysis approach is used for finding the nominal on-chip temperature profile. Both temperature-dependent power and timing analyses are addressed in this chapter.

- The relationships between power, temperature, and timing are illustrated.
- The dynamic timing analysis method (also called delay simulation) is described.
  - It simulates a design with given input vector patterns.
  - It is accurate yet expensive, which is ideal for analyzing small designs.
- The static timing analysis method is described.
  - It finds the critical path timing without requiring the input patterns.
  - Two different approaches are in the static timing analysis method: path-oriented approach and block-oriented approach.
  - The path-oriented approach enumerates the k-most critical paths one at a time.
  - The block-oriented approach propagates the timings through each block and only the worst timing is recorded. In this approach, the details of how to propagating the arrival times, the required arrival times, and the slacks are described.
  - The false path problem in static timing analysis is defined and examined. Methods used to remove the false paths are presented.

- The delay models generally used in timing analysis are introduced.
- A statistical power analysis method, based on MED, is presented. It estimates the power density in a design so that the local temperature rise can be accurately determined.
- A statistical power-temperature iteration scheme for finding the average power and the nominal steady-state power is developed and described.
- The temperature-dependent gate and RC delay models used in timing analysis are described.
- Simulation example are provided. The results show that the on-chip temperature rise and temperature gradient not only can change the circuit timing, but also can change the criticality of a path.

References

[1] D. J. Pilling, and H. B. Sun, "Computer-aided prediction of delays in LSI logic systems," in *Proceedings of the ACM/IEEE Design Automation Workshop,* pp. 182–186, 1973.

[2] M. A. Wold, "Design verification and performance analysis," in *Proceedings of the ACM/IEEE Design Automation Conference,* pp. 264–270, 1978.

[3] R. Kamikawai, M. Yamada, T. Chiba, K. Furumaya, and Y. Tsuchiya, "A critical path delay check system," in *Proceedings of the ACM/IEEE Design Automation Conference,* pp. 118–123, 1981.

[4] R. B. Hitchcock, G. L. Smith, and D. D. Cheng, "Timing analysis of computer hardware," *IBM Journal of Research and Development*, vol. 26, pp. 100–105, Jan. 1982.

[5] J. Ousterhout, "A switch-level timing verifier for digital MOS VLSI," *IEEE Transactions on Computer-Aided Design of Integrated Circuits and Systems*, vol. 4, pp. 336–349, July 1985.

[6] H. C. Yen, D. H. Du, and S. Ghanta, "Efficient algorithms for extracting the k-most critical paths in timing analysis," in *Proceedings of the ACM/IEEE Design Automation Conference*, pp. 649–654, June 1989.

[7] Y. C. Ju and R. A. Saleh, "Incremental techniques for the identification of statically sensitizable critical paths," in *Proceedings of the ACM/IEEE Design Automation Conference*, pp. 541–546, June 1991.

[8] D. H. Du, S. H. Yen, and S. Ghanta, "On the general false path problem in timing analysis," in *Proceedings of the ACM/IEEE Design Automation Conference*, pp. 555–560, June 1989.

[9] J. Benkoski, E. V. Meersch, L. J. Claesen, and H. DeMan, "Timing verification using statically sensitizable paths," *IEEE Transactions on Computer-Aided Design of Integrated Circuits and Systems*, vol. 9, pp. 1073–1084, Oct. 1990.

[10] R. E. Bryant, "Graph-based algorithms for Boolean function manipulation," *IEEE Transactions on Computers*, vol. 35, pp. 677–691, Aug. 1986.

[11] K. S. Brace, R. L. Rudell, and R. E. Bryant, "Efficient implementation of a BDD package," in *Proceedings of the ACM/IEEE Design Automation Conference*, pp. 40–45, June 1990.

[12] P. C. McGeer and R. K. Brayton, *Integrating Functional and Temporal Domains in Logic Design.* Kluwer Academic, New York, 1991.

[13] P. C. McGeer and R. K. Brayton, "Efficient algorithms for computing the longest viable path in a combinational network," in *Proceedings of the ACM/IEEE Design Automation Conference*, pp. 561–567, June 1989.

[14] T. G. Szymanski, "Computing optimal clock schedules," in *Proceedings of the ACM/IEEE Design Automation Conference*, pp. 399–404, 1990.

[15] T. M. Burks, K. A. Sakallah, and T. N. Mudge, "Identification of critical paths in circuits with level-sensitive latches," in *Proceedings of the ACM/IEEE International Conference on Computer-Aided Design*, pp. 137–141, 1992.

[16] T. M. Burks and K. A. Sakallah, "Optimization of critical paths in circuits with level-sensitive latches," in *Proceedings of the ACM/IEEE International Conference on Computer-Aided Design*, pp. 468–473, 1994.

[17] H. Y. Chen and S. Dutta, "A timing model for static CMOS gates," in *Proceedings of the ACM/IEEE International Conference on Computer-Aided Design*, 1989.

[18] T. Sakurai and A. R. Newton, "Delay analysis of series connected MOSFETs," *IEEE Journal of Solid-State Circuits*, vol. 26, pp. 122–131, Feb. 1991.

[19] J. T. Kong and D. Overhauser, "Methods to improve digital MOS macromodel accuracy," *IEEE Transactions on Computer-Aided Design of Integrated Circuits and Systems*, vol. 14, pp. 868–881, July 1995.

[20] A. Nabavi-Lishi and N. C. Rumin, "Inverter models of CMOS gates for supply current and delay evaluation," *IEEE Transactions on Computer-Aided Design of Integrated Circuits and Systems*, pp. 1271–1279, 1994.

[21] S. Z. Sun, D. H. Du, and H. C. Chen, "Efficient timing analysis for CMOS circuits considering data dependent delays," in *Proceedings of the IEEE International Conference on Computer Design*, 1994.

[22] V. Chandramouli and K. A. Sakallah, "Modeling the effects of temporal proximity of input transitions on gate propagation delay and transition time," in *Proceedings of the ACM/IEEE Design Automation Conference*, pp. 617–622, 1996.

[23] M. G. Xakellis and F. N. Najm, "Statistical estimation of the switching activity in digital circuits," in *Proceedings of the ACM/IEEE Design Automation Conference*, pp. 728–733, June 1994.

[24] P. L. Meyer, *Introductory Probability and Statistical Applications*. Addison-Wesley, 1970.

[25] Y. K. Cheng and S. M. Kang, "Temperature-driven power and timing analysis for CMOS VLSI circuits," in *Proceedings of the IEEE International Symposium on Circuits and Systems*, vol. 6, pp. 214–217, May 1999.

[26] Y. H. Shih and S. M. Kang, "Analytic transient solution of general MOS circuit primitives," *IEEE Transactions on Computer-Aided Design of Integrated Circuits and Systems*, vol. 11, pp. 719–731, June 1992.

About the Authors

Dr. Yi-Kan Cheng received the B.S. degree from the National Chiao-Tung University, Taiwan in 1991, the M.S. degree from the University of Southern California in 1993, and the Ph.D degree from the University of Illinois at Urbana-Champaign in 1997, all in electrical engineering. In the summer of 1996, he was with the Technology Computer-Aided Design (TCAD) Department of Intel Corporation, Santa Clara, CA, working in the area of electrothermal reliability simulation and modeling. Currently he is with the Motorola Somerset Design Center, Austin, TX, as a Development Staff Member for the PowerPC microprocessor design. His present research interests include IC design, IC reliability analysis, timing optimization and analysis, and power analysis.

Dr. Ching-Han Tsai received the B.S. degree in electrical engineering from National Taiwan University in 1992, and M.S. and Ph.D. degree in electrical and Computer engineering from the University of Illinois at Urbana-Champaign in 1997 and 2000, respectively. He was with Intel Corp. in the summer of 1997, and Cadence Design Systems Inc. in the summer of 1998. His research interests include electrothermal circuit simulation, substrate modeling for noise/latchup/thermal analysis, and reliability-driven physical design.

Dr. Chin-Chi Teng received the B.S. Eng. degree in electrical engineering from the National Taiwan University, Taiwan, and the M.S. and Ph.D. degrees in electrical and computer engineering in 1993 and 1996, respectively, from the University of Illinois at Urbana-Champaign. Since 1996, he was with the Analysis Product Division, Avant! Corporation, Fremont, CA. Currently he is a senior member of technical staff at Silicon Perspective Corporation, Santa Clara, CA. His research interests are in the areas of computer-aided design on VLSI circuits and systems, with emphasis on circuit simulation, power estimation, interconnect reliability assessment, and post-layout performance optimization for deep-submicron circuits.

Dr. Sung-Mo (Steve) Kang received the Ph.D. degree in electrical engineering from the University of California at Berkeley in 1975. Until 1985 he was with AT&T Bell Laboratories at Murray Hill and Holmdel, and also served as a faculty member of Rutgers University. In 1985, he joined the University of Illinois at Urbana-Champaign where he is Professor and Department Head of Electrical and Computer Engineering, and Research Professor of Coordinated Science Laboratory and Beckman Institute for Advanced Science and Technology. He was the Founding Editor-in-Chief of the IEEE Transactions on Very Large Scale Integration (VLSI) Systems.

Dr. Kang is Fellow of IEEE and AAAS, a Foreign Member of National Academy of Engineering of Korea. He is recipient of the IEEE Millennium Medal (2000), SRC Technical Excellence Award (1999), KBS Award in Science and Technology (1998), IEEE CAS Society Technical Achievement Award (1997), Humboldt Research Award for Senior US Scientists (1996), IEEE Graduate Teaching Technical Field Award (1996), IEEE Circuits and Systems Society Meritorious Service Award (1994), SRC Inventor Recognition Awards (1993, 1996), IEEE CAS Darlington Prize Paper Award (1993), ICCD Best Paper Award (1986) and Myril B. Reed Best Paper Award (1979). He was an IEEE CAS Distinguished Lecturer (1994-1997) and holds six patents, published over 250 papers and co-authored six books, Design Automation For Timing-Driven Layout Synthesis (1992), Hot-Carrier Reliability of MOS VLSI Circuits (1993), Physical Design for Multichip Modules (1994), and Modeling of Electrical Overstress in Integrated Circuits (1994) from Kluwer Academic Publishers, CMOS Digital Circuits: Analysis and Design (1995, 2nd ed. 1998) from McGraw-Hill, and Computer-Aided Design of Optoelectronic Integrated Circuits and Systems (1996) from Prentice Hall.

Index

MIX
Papier aus verantwortungsvollen Quellen
Paper from responsible sources
FSC® C105338
FSC
www.fsc.org

If you have any concerns about our products,
you can contact us on
ProductSafety@springernature.com

In case Publisher is established outside the EU,
the EU authorized representative is:
Springer Nature Customer Service Center GmbH
Europaplatz 3, 69115 Heidelberg, Germany

Printed by Libri Plureos GmbH
in Hamburg, Germany